Digital Planet,

Human Inhabitants

Digital Planet, Human Inhabitants

The Future Was Yesterday

Practical Tips to Future-Proof Your Career, And ~~Survive~~ ~~Thrive~~ - Thurvive - In the Early 21st Century

Kartik Sakthivel PhD

Published by:
Brain Fuel Books LLC
PO Box 413
Dover, NH 03820

Interior Design: Kartik Sakthivel
Cover Design: Leslie Sano

Paperback ISBN: 978-1-7377167-4-7
Hardcover ISBN: 978-1-7377167-5-4
Digital ISBN: 978-1-7377167-3-0

ACKNOWLEDGEMENTS

To my parents, P.R. Sakthivel, and Dr. Veena Sakthivel, without whom none of the words in this book could have been possible.

With many thanks to my family; Kiko, Sophie, and Simon, who put up with me being unavailable nights and weekends because I was sequestered in my office working on this book.

To the millions of workers who are on the cusp of a technological revolution – I am rooting for each of you to capitalize on this era and prosper like no generations have before – for the betterment of all humankind.

Table of Contents

FOREWORD (partially authored by AI)

The pace of technological change has never been faster, and it shows no signs of slowing down. As we hurtle towards a future shaped by artificial intelligence, automation, and other technological advancements, it's clear that technology will play an increasingly central role in our daily lives. As we look ahead to the next 50 years, it's clear that technological change will continue to shape our world in profound and unprecedented ways. From artificial intelligence and automation, to virtual reality and beyond, the technologies of the future will have a significant impact on how we live and work.

For many of us, this rapid change can be intimidating and even overwhelming. But it's important to remember that technology is not something to fear, but rather a powerful tool that can help us solve some of the biggest challenges facing humanity. That's why it's more important than ever for humans to adopt and adapt to new technologies. By staying open to new ideas and learning about the technologies of the future, we can not only stay relevant in a rapidly changing world, but also shape the direction of technological development in a way that reflects our values and goals. By understanding the technology that surrounds us, we can not only better navigate this rapidly changing world, but also shape it in a way that reflects our values and goals.

This book is a guide for those who want to embrace the future and take control of their own careers. Through engaging and accessible writing, the author explores the most important technological advancements shaping our world today, and offer practical guidance on how to use them to your advantage.

So if you're ready to face the future with hope and optimism, this book is the perfect place to start. By adopting and adapting to new technologies, we can shape a brighter future for ourselves and for generations to come.

PREFACE

What if I told you that I am 100% confident that at least 50% of the tasks you perform today as part of your job will be automated in the next one to two decades? What if I told you that depending on your profession, there is also a greater than 50% chance that 100% of your job will be automated in the next one to two decades?

We are in unchartered waters with all the adventures and perils that unchartered waters can bring. This era is exciting and presents an opportunity to those who can catch the rising crest, but can prove to be a treacherous and perilous journey to those who choose to coast along. We are living through unprecedented times, and unprecedented times present an incredible opportunity to thrive, not just survive amidst events that are well outside our control. Change can create challenges, but change also creates enormous opportunity. This book will present you with practical tips and tricks that you can leverage to navigate these unchartered waters. The guidance provided will hopefully help you to take command of your trajectory and career prospects to ensure you can capitalize on change.

Following the recommendations outlined in this book, you will set yourself up for success to survive our own mindset around the pace of change that is outside our control, and thrive amidst the abundance of opportunities that this digital era will create. We will learn how to survive against our own hardwired bias against change, and learn how to thrive in this era – we will ready ourselves to "**thurvive**" this age of unrestrained and accelerating digitization. This book will follow four sections. First, we will seek to ***Understand the Change*** - The first step is to recognize that the technology-driven changes are an inevitability. We will take a step into the future in the first section of this book and took a peek at the most conservative likely look into our future. Next, we will ***Understand the***

Need to Have to Change - The second step is to understand that we do not have a choice to slow, stymie, or stop these technology-driven changes. In section two of this book, we will explore the past and seek to ground ourselves on the fact that while the past offers some prologue to the future, the increasing and accelerating pace of change over the past century is a harbinger on shape of things to come. The third section of this book will help us ***Understand the Need to Want to Change*** – The third step is that we ensure we are proactively addressing change readiness. In section three, we will return to the present after having explored the future and peeked at the past. We will examine how the recent past is influencing our world today and what that might portend for our occupations over the next few decades. Finally, we will ***Understand How to be Ready for the Change*** – The fourth step is to adopt the recommendations outlined in this section, building your own personal plan on the foundation of a continual learning mindset and keep exercising that continual learning mindset. This section will also present the ***Four Techniques for Thurvival*** in this digital era.

When the future is uncertain, anything is possible. This book will prepare you for that digital anything.

SECTION ONE

UNDERSTAND THE CHANGE

Our Digital Future

Digital Planet Human Inhabitants

Chapter One – "Today" is the "Tomorrow" you talked about "Yesterday"

Come with me as we embark on a journey of imagination, one as fantastical as you would like it to be. The limits of where we will travel is only limited by the confines of our own imagination. However, since every journey needs a destination, let's imagine ourselves in the future, just twenty-five years from now.

The year is 2046. It is the end of your workday, and you are getting ready to return to your home in the exurbs. The nearly 150 mile commute, which once would have been unimaginable, has provided you (and dozens of your neighbors in the "Exurbia Prime Subdivision") with the opportunity to own "more home for the price" in the immediate suburbs, than you otherwise would have been able to. You physically go into the office once a week, on Wednesdays, primarily for the socialization aspect of being part of an organizational culture. You recall a time early on in your career when you spent more than two hours on a back-and-forth commute from your condominium in the suburbs to your office building. Battling traffic at times, and attempting to sync your commute with public transport schedules at others, you reflect on the countless hours you spent in the stressful, mundane, tedious, and tiring process of just getting to work and back. "A young person's life", you think to yourself.

You remember the COVID-19 pandemic and how the world virtualized through that historical event. Emerging from this event and starting in 2022, you reflect upon how the world's collective attitude, including your own, shifted significantly on what it means to "get to work". For over two years starting in 2020, tens of millions of workers across the world successfully worked from home. In the aftermath of the pandemic, firms and workers alike realized that their productivity skyrocketed and remained consistently high in the new work culture. Employees reported being happier, having achieved a better work-life balance, and they reinvested the time and energy they had spent in commuting in family and personal wellbeing. Most workers across industries, having realized that remote/virtual work was a viable option for them, chose to work in a "*hyperflex*" or hybridized model.

With millions of unfilled jobs, and workers considering changing jobs and companies in pursuit of better remote work options, subsequent years saw organizations across industries embrace virtual work, hybrid working models, and the concepts of "hyper-flexibility" as throngs of workers embraced a better work-life balance. Once the "new normal" settled in, companies urged their employees to take vacation time and regular time-off to strike a good balance between productivity and prevention of employee burnout. It was a seminal moment for the world and proved to be a lasting cultural shift in how people thought about work and life.

The thought of home reminds you that today, Wednesday, is lawn-mowing day. With the exurbs allowing you to afford a larger house on a five-acre plot of land, there is a considerable amount of lawn to mow. You flick your wrist, and the familiar logo of an apple with a single bite out of it that adorns your svelte wristband lights up, projecting your personal digital assistant into the air above your palm. You swipe at the air to navigate through the screens and click an "air button" to activate your smart-home app. Scrolling through the app, you check the live feed of the front-facing external camera of your house and see an impeccable, freshly mowed lawn. For a ten-year old Artificial Intelligence (AI)-powered robotic lawn mower, your Husqvarna® continues to do an excellent job. The *connected* AI-powered robot operates on a weekly schedule, programming itself to mow and mulch your acre of lawn within your five acre property. The mower is smart enough to know what days and times of the week rain is in the forecast and to skip those days and times.

You flick your wrist again to disable the projection and resume packing up. The frameless television behind you presents a commercial for the latest model of the Husqvarna® AI-powered robotic lawn mower, with the CNBC business program that was on appearing in a box on the side. It captivates your attention and you pause to watch the entire commercial. The newest model has built-in soil-sensor detection capabilities. Sampling your lawn as it mows and immediately analyzing the soil conditions – pH levels/acidity, minerals, moisture, etc. – the latest model treats the soil with organic lawn treatments immediately after mowing is complete (the commercial calls this process "the Triple T: tasting-testing-treating"). If the ground is determined to be arid, this mower model has the capability to communicate with your "smart-home" and adjust the amount of water used by your sprinkler system to water the lawn. Smart home systems via home automation (called domotics) began their rapid development through the 201xs and continued to evolve at a breakneck pace. All new home construction starting in the 2030s involved some level of domotics (Hill, 2017), which is building automation for a home, called a smart home or smart house to monitor and/or control home attributes such as lighting, climate, entertainment systems, and appliances, home security such as access control and alarm systems. Having been plugged into the internet, home devices became an integral part of the Internet of Things ("IoT").

Impressively, this new AI mower robot has the ability to adjust itself seasonally, regulating the precise mix of lawn treatment compounds based on changing weather conditions, viability of the soil, the growth zone your home is in, and even the type of grass for that particular zone, with those specific climatic conditions. The commercial says that this is similar to the technology that was first used by the NASA Mars rovers as they looked for organic compounds in the soil of Mars in the 2020s before manned missions to establish human colonies on the "Red Planet". Little did we know in the early 21st century that as we looked for microbial life on Mars, the Jovian moons Europa (NASA, n.d.) and Ganymede - and Saturn's moon

Enceladus (Wall, 2021) – and potentially Titan - were teeming with aquatic life under their icy crusts, just as in 2035, Google's Quantum Computing-powered AI algorithms (Google, n.d.) had predicted that they would be.

A handy feature of the new Husqvarna®, the narrator says as a bounding Golden Retriever dog leaps for a tennis ball on a gorgeous green lawn, is that the mower is smart enough to detect and dispose of pet excrement before it mows your law – a great feature for pet owners. "Maybe I will get this for myself as a birthday present", you tell yourself as a commercial for a competing AI-powered mower airs by rival John Deere®. You dismiss the ads by standing in front of the TV projection and swiping left. You remember that your neighbors, the Joneses, purchased the same new John Deere® model a couple of months ago and could not stop raving about it. "Jones, I can't stand that guy - showing off all the time - but his wife and kids are so sweet", you think as you use your thumb to gently press the underside of your wedding ring.

"Alexa", you speak into your ring as the silicone band lights up in a luminescent blue, "get me a ride-share please". "Would the location be to Northside Park and Ride, Row K, Port 12?" your ring asks. With your *smart-ring* (and other connected devices) having an in-depth knowledge of your travel patterns, you rarely have to correct your destination. "Yes, the Park and Ride" you respond. "Sure," your smart-ring responds, "please proceed to Vertiport S at your location on the South Tower. Estimated time of arrival is 7 minutes." While you could have well asked for a ride-share via your smartphone wristband, you prefer to designate specific AI assistants for specific tasks. It's just your personal preference going back to when you were a child in the early part of the 21st century.

As one of the first waves of the generation termed as "Generation Z", you were the first generation that was used to having smartphones since Elementary School. In fact, your generation was the first fully digitally enabled generation, connected to the internet via your smartphones at all times. You also used several other devices – smartwatches, separate gaming platforms, physical smart TVs for streaming media, holographic platforms, etc. – all offering some level of overlapping functionality, and yet your old habit of device specialization (using specific devices for specific things) has not been easy to shake. The consolidation of devices accelerated in the 2030s. Affordable AI was available within all consumer electronics, creating a connected digital world, where devices could communicate with each other (as compared to being only able to communicate with their human owners) for the purpose of providing better service to their owners.

Unlike you, most people were shifting their preferences away from multiple, specialized devices to single devices/AI-powered personal digital assistants. You are fully aware that in the next twelve-to-eighteen months, the commercialization of embedded AI *nanotech* devices will begin to render peripherals - things such as your wristband and your ring - obsolete. After all, it was about fifteen years earlier when these devices rendered computer peripherals such as keyboards and mice obsolete. As a computer science graduate, you chuckle because inventor and futurist Ray Kurzweil's prediction of the year when humankind would achieve "Technology Singularity" has been eerily accurate (Reddy, 2017). Kurzweil made almost 200 futuristic predictions through his career and had an impressive accuracy rate of close to 90%. He defined 2045 as the year "Singularity" would be achieved by defining this moment as - "which is when we will multiply our effective intelligence a billion-fold by merging with the intelligence we have created." (Reddy, 2017).

In other words, Kurzweil had accurately predicted that by 2045, humans would integrate intelligent technology *within* our bodies – or in sci-fi terms, merge with intelligent machines – to boost our intelligence and dramatically improve our quality of life. Most major technology providers had recently introduced implantable devices across North America and Western Europe. Asian countries such as Japan, China, and India had gotten a bit of a head start with AI devices directly implanted inside humans. These consumer electronics type devices, embedded into the epidermal layers of the skin, effectively merged humans with AI-powered machines, eliminating the need to purchase peripheral equipment such as smart-wristbands and smart-rings. The rollouts had predictably been met with protests stemming from online conspiracy theories the clamored of corporations, and secret societies with nefarious agendas, and "the government" using AI to control and subjugate their population. Additionally, the requisite scrutiny and approvals from US Federal Agencies such as the Food and Drug Administration (FDA), and these implantable computers being embroiled in US Healthcare System over the debate of whether purchase of these devices constituted an elective medical procedure, caused some delays in a widespread rollout. In fact, you remind yourself that computers have been implanted in human's brains since at least the 2020s and had made a tangible difference in recipients such as patients with Parkinson's disease and those dealing with memory loss issues.

"Everyone likes to protest something" you mutter under your breath, taking umbrage to the fact that people forget that in addition to computer chips that were implanted for very specific purposes and to treat very specific conditions, medical *nanobots* had been injected into patients for the better part of the past decade. Nanobots, as the name suggests, are minute robots (bots) - blood-cell sized devices that are introduced directly into the human body, allowing a host body to combat diseases as well as to improve memory and cognitive abilities. Just like Ray Kurzweil had predicted, nanobot technology successfully completed clinical trials and went mainstream in the mid-2030s. With an initial implementation sponsored by the United States Department of Defense – the Defense Advanced Research Projects Agency (DARPA) – US military personnel deployed in tropical countries were administered medical nanobots in the early 2030s. Having proven highly effective in immunization against tropical diseases, these nanobots had also allowed the military to monitor, in real-time, the health of each soldier in the field. Not without controversy, medical nanobots for military use had dealt with their own set of conspiracy theories about the military developing "super soldiers". The reality was far more benevolent.

Nanobots provided thousands of bits of information about a host soldier back to central command, including, but not limited to such as aspects as - blood pressure, heart rate, oxygen absorption, and mineral and vitamin deficiencies. When coupled with Quantum Computing powered AI, the nanobots also provided a mechanism to predict diseases such as diabetes and some forms of cancer. When nanobots proved to be a resounding success in the field with Earth-based military personnel, their introduction in the space-faring community was astonishingly swift. Nanobots became commonplace for astronauts (and cosmonauts) working for private space agencies such as SpaceX and Blue Origin, in addition to governmental space agencies such as NASA, the European Space Agency (ESA), Russia's Roscomos, China's CNSA, and India's ISRO.

The mainstream implications were profound and medical nanobots had been transforming medicine and preventive care on a scale not seen since the discovery of penicillin (and ergo, antibiotics) in 1928. With initial nanobot implementations allowing for patients with Alzheimer's disease to maintain their cognitive functions, nanobots were highly helpful in detection and treatment of early onset dementia patients. Trials where nanobots had accurately detected fetal development issues and leveraged AI to prescribe recommended courses of action had been remarkably transformative in saving countless lives – of infants and their mothers. The infusion of technology into humans had already dramatically improved the quality of life for millions, with additional advancements promising to finally find an AI-led cure to some of humanity's persistent diseases such as diabetes, cancer, and other genetic maladies. Medical nanobots were as transformative as gene-editing technique CRISPR proved to be in the late 2020s and beyond, and probably just as controversial and susceptible to fact-free conjecture and conspiracy theories. Twenty years earlier, you had first read a news article about CRISPR that appeared in the "New Scientist" and the promise of this technology to transform the world - "CRISPR is a technology that can be used to edit genes and, as such, will likely change the world.

The essence of CRISPR is simple: it's a way of finding a specific bit of DNA inside a cell. After that, the next step in CRISPR gene editing is usually to alter that piece of DNA. However, CRISPR has also been adapted to do other things too, such as turning genes on or off without altering their sequence." (Scientist, 2021). "CRISPR" (pronounced "crisper") stands for Clustered Regularly Interspaced Short Palindromic Repeats, which are the hallmark of a bacterial defense system that forms the basis for CRISPR-Cas9 genome editing technology (Broad Institute, 2021). CRISPR, which allowed scientists to make specific and precise changes in human DNA, and had been credited with treating countless patients and hundreds of genetic diseases, had been the first to infuse billions of microscopic structures known as nanoparticles into humans. Cells were taken from a patient, edited, infused with nanoparticles that carried genetic instructions, and reintroduced back into the patient (Stein, 2021). Setting aside conspiracies that CRISPR would lead to the wealthy choosing to create "designer babies" and controversies such as with proposals of using CRISPR to genetically resurrect the extinct Wooly Mammoth (Neuman, 2021), the actual medical results had been remarkable across the board (NPR, 2021).

You are about to leave your office and head for the elevators to the "Vertiport" when your smart wristband glows and pulsates, indicating an incoming video call, the "Answer" and "Decline" buttons now hovering in the air, suspended a few inches above your wristband. The Caller ID indicates that it is your physician's office. You stop in the vestibule and click the "Answer" air button. A holographic projection of your physician's receptionist appears. She says she is calling to confirm your appointment for your annual physical exam. You almost start to exchange pleasantries, but remember that you are not talking to a real person, but with an AI holographic projection of your doctor's scheduling software.

The "AI avatar" that your physician happened to choose was a "digital twin" of her assistant. The holographic AI assistant informs you that she already has most of your vitals because your "smart clothes" have already transmitted this data. Your smart wristband provided additional medical data points that could likely make the need for them to draw blood for testing (as a routine procedure) unnecessary. Finally, she reminds you that the sensor in the left shoe of your favorite pair of sneakers is malfunctioning, reiterating how important it is for you to maintain the appropriate weight on your left foot as part of your rehab after having broken your ankle last winter while skiing in Aspen, Colorado. You never fail to be impressed with the clinical yet empathetic manner in which these AI avatars attempt to simulate the human emotional range. While AI has not yet mastered or replicated the range of human emotion, there has been significant improvement in just 20 years. You confirm your appointment, and end the call, assuring the AI avatar that you will get your shoe sensor repaired at the earliest. Never did you think while going through school for Computer Science decades ago, that you would be nagged by a robot.

Telemedicine emerged as a viable option for most basic doctor visits during the COVID-19 pandemic through 2020 and into early 2021. Before vaccines were able to turn the tide of the pandemic, most patients turned to telemedicine as a standard practice, conferring and consulting with their physician over video conferencing platforms such as Zoom, Microsoft Teams, Google Meet, and Cisco WebEx. For many months, doctor consultations were possible only on virtual platforms, something that had never yet been considered seriously as a viable option, and eventually led to a marked change in attitudes towards virtual medical consultations. In the decade after the pandemic, telemedicine rapidly evolved to include medical diagnostic analyses being performed remotely by doctors, aided by connected wearables from smartwatches to connected apparel. The emergence of embedded devices and *nanobots* over the past few years are proving to be a game changer, you surmise, as a physician can access patient vitals from within the patient's body, instead of from external sources such as wearable devices and apparel.

"Lights off", you say into a nondescript panel as you exit your office. Your office lights would have turned off automatically within 90 seconds of the room not detecting motion and a bio-signature within, yet it is pure force of (good) habit that your parents instilled in you to make sure you conserved energy at every opportunity. You think of all the technological advances that are so commonplace now that you tend to take them for granted. How incredibly futuristic these innovations would have seemed at the turn of the century. You remember your father and mother, both accomplished professionals in their own fields, advising you and your siblings to instill in you a mindset of intellectual curiosity and a culture of continual learning. "The future", they insisted, "will be here before you know it". "Change will be upon you whether you want it or not, whether you are ready for it or not. Only continual invention and reinvention will help you succeed in your profession", they advised. You pause to think of how invaluable that advice has proven now when you can reflect on it in your 40s.

As a teenager, their guidance seemed a bit naïve, at times far-fetched, and sometimes even cringe-worthy when you had friends over – you sarcastically called it the "The Gospel of the Digital Millennial". But you sure are glad you listened. A favorite "dad-ism" that you and your siblings often poked fun at was his favorite saying, "'Today' is the 'Tomorrow' you talked about 'Yesterday'". That was his way of conveying that the future that most people envisioned is already upon us. Your mom always retorted with "'Yesterday' was the 'Tomorrow' that we talked about 'Yesterday'" – her way of underscoring her opinion that professionals across industries cannot underestimate the rapidly accelerating pace of change. "If you stop too long to think about it and mire yourself in analysis-paralysis, the change will run roughshod over you". "The Future was Yesterday", your dad used to respond – always eager to subtly sneak in the last word. Your mother was absolutely correct.

As you begin your imaginary journey home, our journey of imagination is only starting. As the future "you", twenty-five years from now, packs up your office and continues on your journey home, so shall we continue our journey of imagination into 2048.

Chapter Two – The Future Was Yesterday

You continue walking down the aisle through the multiple security controlled doors - that open and close automatically as you walk by - into the vestibule towards the elevators. With cameras embedded throughout the building, capturing and analyzing everything from your posture to biometric data, no employee needs to carry a physical access card or use retinal or fingerprint biometric scans for access. The entire building is aware of where every employee is at any given point in time, eliminating the need for additional security checks. You enter an elevator with two of your colleagues, making small talk about last night's football game. Each of them is disembarking at two different Vertiports on floors 52 and 54, as you proceed to your designated pickup location - Vertiport S on floor 57.

Vertiports gained popularity in the mid-2030s and experienced explosive growth, becoming increasingly ubiquitous across every urban landscape. A cross between a helipad and sheltered bus stop, Vertiports, short for "Vertical Ports", were designed as "air taxi stands" for eVTOL (Electric Vertical Takeoff And Landing) Air Taxis. Most major rideshare providers such as Uber® and Lyft® operated massive networks of pilotless (unmanned), AI-controlled and flown, all-electric air taxis throughout metropolitan areas around the world. Vertiports, shared by multiple rideshare providers, were constructed at several public locations across cities, and also conveniently located in most major office buildings. It became common for buildings over 30 stories tall to have multiple Vertiports. Remarkably, in their near 15 to 20 years of continued growth and expansion, and of ferrying millions of commuters, air taxis had enjoyed a perfect accident-free record and no human fatalities. You recall a time in the middle of the 2020s when eVTOLs seemingly mushroomed out of nowhere and became dramatically popular across the world.

The early 2020s saw major automobile manufacturers such as Toyota partner with Uber and Joby Aviation in California, USA (Chapman, 2021) and Japan Airlines (JAL), and rival All Nippon Airways (ANA) partner with the German company Volocopter (Nikkei, 2021) to bring air taxis into service. At the outset, Uber aimed to have driver-operated air taxis ferrying travelers across cities in California in an environmentally-friendly manner to ease the dense traffic congestion, especially invaluable on Californian highways with legendary congestion such as the "405". Air taxis would eventually be available as vehicle options for travelers across most major ridesharing platforms. In Japan, JAL and ANA, accurately predicted that the eVTOL sector would become a trillion-dollar industry over the subsequent decades, and sought initially to have driver-operated flying taxis bring passengers from major airports to the city center across cities in Japan's Mie Prefecture (Pande, 2021).

In-person (actual) drivers eventually gave way to remote-drivers who operated these air taxis akin to remotely flying drones before the pilotless/driverless/unmanned generation of eVTOL craft took to urban skies. A form of these driverless craft/autonomous vehicles existed in service since the early 2020s in Dubai, United Arab Emirates, but faced scrutiny and regulatory hurdles in countries such as the United States before becoming commonplace. Leading the worldwide charge was California, which in turn was led by the city of San Francisco that banned all fossil fuel operated vehicles within its city limits in the mid to late 2030s. This ban was quickly followed up by a phase-out ban on all driver-operated vehicles in favor of a 100% "self-driving vehicle city". Ironically, the ban on driver-operated vehicles in San Francisco had little to do with air taxis, and everything to do with promoting ground-based self-driving electric vehicles. Until that San Francisco ruling, self-driving cars faced continual hurdles in widespread adoption due to concerns about safety, but not the concerns one might have expected. Self-driving cars, despite some early setbacks, had proved to be notably safe in every National Highway Traffic Safety

Administration (NHTSA) controlled test. That was until the predictability of AI-driven autonomous vehicles had to, in the "real world", cohabit and share the roadways with the most unpredictable variable of all – human drivers and driver-operated motor vehicles. Most traffic accidents involving fully autonomous self-driving cars always involved a human driver as the other party/parties. Cities found it difficult to prevent motor vehicle accidents and although there were no fatalities, it was clear that even the smartest AI-powered vehicles lacked the ability to predict the unpredictability of concurrent multiple human drivers sharing the motorways.

In contrast, air taxis and flying cars in general were considered to be "greenfield" and a metaphorical blank canvas. With no precedent, all that was required was the establishment of some basic rules: AI-governed controls on how high and fast flying cars could travel; the requirement that each flying car, regardless of manufacturer (Toyota/Tesla/Joby) or operator (such as Uber/Lyft) could communicate with every other vehicle using a common standard, such that they could map out the travel pathways of every other craft in the sky; avoidance to other aircraft such as helicopters, drones, and airplanes; and comply with the latest Federal Aviation Administration (FAA) guidelines and regulations. When these vehicles switched from driver-operated to self-driving/autonomous mode, it further eased the regulatory requirements on companies that mandated drivers be licensed for operating Light Aircraft Vehicles (a new driver's license class for motor operators that was introduced in the late 2020s).

The implicit promotion in San Francisco of self-driving/self-piloted/autonomous vehicles was a boon to the air taxi industry, especially self-driving/autonomous eVTOL vehicles. Air taxis had proved invaluable to metropolitan areas and also to smaller cities. Cities such as San Francisco and New York bounced back relatively quickly after the COVID-19 pandemic and subsequent normalization of full-time remote or hybrid work. With the AI technology hubs in San Francisco and the financial hubs in New York City, it was not too difficult to attract a new generation of young urban professionals to these metros. Smaller cities however, had to invest in revitalizing themselves. Individuals with families, or those seeking a more laidback lifestyle (like yourself) had decided to leave the smaller metros, choosing to move into the suburbs and exurbs and being fulltime remote workers. Everything from cafes and restaurants that relied on the patronage of office workers during the lunch rush, to custodial services and gyms were inflicted with a precipitous and sometimes fatal loss of business. Entire office buildings went unoccupied or "under-occupied" for years as throngs of workers embraced virtual work. These cities struggled to maintain their civic budgets as tax collection fell because of the loss of businesses due to virtualization.

The decline in the value of commercial real estate had reverberations through local economies. Several of these affected urban areas went through a renaissance of sorts in the early 2030s. Reinventing themselves as "smart" cities, they transformed into technologically progressive cultural hubs. Starting with San Francisco, the change moved across other cities in "The Golden State" (California History - State Symbols, n.d.); soon, taking cues from California, several cities across multiple states banned personal vehicles and gasoline-powered automobiles (modeled after Sweden that switched 100% to electric vehicles in 2035), and some even banned driver-operated vehicles altogether. They invested in "greening" themselves by easing traffic and congestion, directing their budgets into infrastructure, and attracting large corporations from big cities to set up "annex" offices in their locations. Assisting the transit between the larger cities to these smaller cities was a rich network of air taxis. Cutting travel time into a fraction of the earlier commute between bigger and smaller cities, such as Boston/New York City to Hartford, Connecticut, corporations were able to have their cake and eat it too by maintaining a presence in metros like New York, while doing a large portion of their work from smaller cities and towns to reduce their operational costs.

Attracting more young urban professionals than ever before, cities were dominated by activities and lifestyles that appealed to these young, single urban professionals. Riding the elevator with your colleagues you look outside the glass façade of the rapidly moving elevator at some of these young, urban professionals on each floor as the elevator ascends. As a Generation Z born professional, you feel that these coworkers, belonging to Generation Alpha (Gen A) – those born between 2012 and 2025 – are a remarkably different generation from yours, yours being the first digital generation. After all, Gen A grew up with AI-enabled consumer electronics, the advent of quantum computing, flying cars, and embedded nanotechnology. They are literally and figuratively wired very differently, you think to yourself.

"Gosh, the new employees seem to be getting younger" you remark to your peers riding up the elevator with you. "They aren't getting any younger, you're getting older", one of your peers retorts with a wry smile before disembarking at their floor. "Aren't we all?" you wonder aloud as your peer disembarks from the elevator. Like every generation says of the next one, you tell yourself that this generation has it much easier than you did. When you entered the software engineering profession, specializing in the burgeoning data science and analytics field, you were required to learn and become adept at several programming languages. Despite the availability of a multitude of tools and online assistance, where you could swiftly find answers to most programming questions and code snippets, knowledge of how to author code – elegant, reusable, secure code – was a must-have. Rising through the leadership ranks in the profession required you to build on this bedrock of technical competency with a lopsidedly heavy use of your soft-skills. You had found that some of your highly technically skilled peers back then were lacking in their soft-skill development. "A generation that prefers to text rather than to talk" you recall your dad saying of your generation.

As a second-generation professional in the software engineering sector, you marvel at the thought that when your father broke into the software development field, he had no access to the kind of tools that you had when you began. "Books", you recall your dad saying, "…we had books. We used to have to look up each and every software code function and "method". Then compile a program. Watch it fail. Fix. Run again. It was time consuming and tedious. You have it so much easier." There was probably something enriching about having to have innate mastery of a programming language and through trial-and-error, assemble a complex program working alongside several other engineers with relatively the same advantages and challenges as you. Your father's generation experienced the same profession very differently than did your own generation. With the widespread availability of prebuilt software libraries, open source software, and an API-driven (Application Programming Interface) ecosystem, technology and software development as a profession became relatively much easier to practice – the "democratization of software", as your father referred to it.

A pivotal new era commenced around 2021 with the advent of AI-authored code. Although rudimentary at the time, the advent of OpenAI Codex (OpenAI, 2021) proved transformational on how software was written. In 2016, OpenAI Codex was launched as a platform that translated natural language (such as English statements) into software code. OpenAI Codex - built by OpenAI - a company cofounded by Elon Musk (amongst others), was, at the time, the world's most ambitious artificial intelligence research laboratory consisting of the for-profit corporation OpenAI LP and its parent company, the non-profit OpenAI Inc. The company was said to have been lionized for its mission. Its goal was to be the first to create Artificial General Intelligence, or AGI - a machine with the learning and reasoning powers of a human mind. The lab's goal was to ensure that AI technology was developed safely and its benefits distributed evenly to the world.

The OpenAI charter declared that OpenAI's "primary fiduciary duty is to humanity." (Klok, 2020). AI-authored software products necessitated a new way of programming for those entering the computer science field. As opposed to having to author code for a complex program from scratch, engineers could focus on either tweaking or fine-tuning AI models, or helping AI learn new things. A burgeoning field of study within computer science, in addition to cybersecurity and data science, was within the ethical AI field. The ongoing mission of ethical AI was to ensure AI algorithms continued to benefit humanity and were free from the early days of "Narrow AI", where "clumsy AI" used to be inadvertently biased and fragile. You recall that during these early days of "Narrow AI", AI models could be used to perpetrate great abuse and deception. Since the expense of developing and running these rudimentary models was high, their power was concentrated in the hands of a few. Companies such as OpenAI feared that, by extrapolation, AGI could be catastrophic without the careful guidance of a "benevolent shepherd", which they aspired to be (Klok, 2020).

In what has been a fascinating shift the computer science landscape through the 2030s was to witness, the focus would be less on "Machine Learning" (ML), and more on the flipside of that coin, the aspect of "Human Teaching". Human programmers were trained to make smart AI - smarter, and smarter AI in turn led to humans having to learn new ways of making themselves even smarter. A symbiotic relationship, which, proving Kurzweil correct yet again, around 2030, computers surpassed the computational capacity of what was once the smartest and most complex computational force on this planet - the human brain (McSweeney, 2020).

"Vertiport S", the elevator AI voice reports as the double doors open. You step out of the elevator and head to the Vertiport shelter where you are greeted by three other fellow employees who will be traveling to the Northside Park and Ride with you in the same air taxi. You wave, recognizing your colleague Bob from the Accounting department. "Bob from Accounting", looking in your direction through his sleek Apple Smart Glasses, waves back at you and raises his index finger, indicating that he will be with you momentarily. "Ah, he's either reading the stock ticker or the news" you tell yourself. The newer smart glasses are "smart" enough to have their front-facing cameras detect gestures (or incoming obstacles/dangers) directed at the wearer and pause the program the wearer is viewing in order to give them an opportunity to evaluate and respond. You smirk because smart glasses were probably one of the only AI-enabled consumer electronics products you personally did not own and never have. Having spent a good sum of money for LASIK vision correction so that you did not need to wear glasses, the last thing you wanted to do was trade vision correcting smart glasses for mere smart glasses, or even smart contact lenses.

After significant hype for a few years starting in 2012 with the Google Glass, smart glasses had experienced what the research firm Gartner referred to as the "Trough of Disillusionment" part of the technology adoption curve they refer to as the "Gartner Hype Cycle". Gartner Hype Cycles provide a graphic representation of the maturity and adoption of technologies and applications, and how they are potentially relevant to solving real business problems and exploiting new opportunities. (Gartner, n.d.). Smart glasses, starting with the "failure" of the Google Glass, were relegated for nearly a decade to specialized manufacturing, design, gaming, and augmented/virtual reality, in stark contrast to the initial idea that these devices would be leveraged for everyday use. "Google originally advertised Glass in terms of experience augmentation. A Google Glass advertisement in 2012 featured skydiving, biking, as well as wall scaling, and eventually, the videos showed user-friendly information instantaneously appearing on-screen during everyday activities. Google's aspirations were lofty - the technology required lengthy battery life, improved image-recognition capabilities, and a lot of data. Rather than augment reality, Glass simply supplemented it. The three-to-five-hour battery life enabled users to check messages, view photos, and surf the internet. However, Glass was competing with other devices that boasted superior cameras, larger capacities, and faster processors" (Weidner, 2021), and sales did not match expectations.

The smart glass wearable consumer electronics market experienced a renaissance in the 2030s after following a long, strange, winding journey. As it turns out, to make smart glasses successful, consumers first needed to become used to "smart glass" – it was less about glasses and more about the applications of internet-enabled panes of glass in everyday goods that made consumers comfortable and habituated to the technology. Starting with Augmented Reality/Virtual Reality headsets that provided the gaming community with an immersive video gaming experience and quickly becoming immensely popular within the adult entertainment industry (for similar immersive experiences), smart glass technology expanded into widespread commercial use. Finding its way into everything from the dashboard heads-up display for the inside of an automobile windshield to ski helmets, smart glass was everywhere seemingly overnight. Rideshare companies used the technology on their vehicles to run advertisements to passengers, who could be exposed to an endless stream of commercials while looking out the window. The smart glass found its way into helmets from astronauts to motorcyclists, proving indispensable for space travel as well as to riders with real-time information on the road conditions including potholes and obstacles on the motorways and weather conditions,. This resulted in marked decrease in motorcycle fatalities since riders could anticipate and avoid unpleasant surprises on the road. Once smart glass became seemingly ubiquitous – from the turnstile of a building's revolving door to your bathroom mirror – the technology re-emerged as a wearable platform similar to the Google Glass. "A technology before its time" claimed Alphabet, the parent company of Google, salvaging a rare early failure in the history of their vastly successful technology conglomerate.

"How's it going?!" - Bob greets you with his usual exuberance. "Way too animated for a guy from accounting", you think to yourself. "Good, but that can change - how is the market today?" you ask cautiously, "…that is presuming you were reviewing your stock portfolio", you gesture with your finger to your face where a pair of smart glasses would sit. "Oh, sorry about that, I was at a critical juncture in my immersive book", Bob responds. Immersive books had been the rage for at least the last ten years. Starting in the mid-30s, AI-authored books became very well-written (and popular as a result) that the New York Times had made the highly controversial decision to feature them on the bestsellers list. Naturally this had outraged the author community who saw writing books, regardless of genre, as a fundamental human expression of thought and art.

AI-authored books grew to be so well-written that they eventually started competing for top of the New York Times Bestsellers lists and some even managed to outsell human authors. AI-books, over a decade, expanded to become fully immersive experiences and spawned a whole new genre of literature. Readers could now fully implant themselves into the book's plot and the AI would randomize and generate a whole new sequence of events, each time eventually leading to the final outcome of the book. For a reader, this meant an incalculably large number of permutations and combinations to experience the same book, ultimately arriving at the same outcome. For nonfiction, it allowed readers to pause at certain exercises or lessons and attempt to role-play or "live" out real-life scenarios that the book espoused. "Oh, here's our ride!" exclaims Bob, once again making you cynically wonder whether his propensity for being excitable was a side-effect of some supercharged nanobots functioning within him.

You see your rideshare arriving as it makes its final approach onto the Vertiport. At this time of the day, the 8-seater Toyota is empty – a function of not having passengers travel inbound to the city's downtown area at this time of the day. The air-taxi makes a seamless and smooth landing and its wing doors swing open. Bob, another person you do not recognize, and you make your way into the cabin and get comfortable in the bucket seats while the automated seatbelt fasteners strap you in for your journey to the Park and Ride. Bob waves at you as the privacy partitioning between each passenger rises up and the pre-flight safety instructions are narrated by the pilotless AI-operated vehicle. With an almost imperceptible whoosh, which you've privately thought of as "Superman-ning it", the Toyota eVTOL vehicle takes off vertically into the air. The private screen within your now-partitioned private pod displays your flight time, duration to destination, and displays a map that shows other rideshare flights in the area. The map is bejeweled with eVTOL rideshares, ranging from Uber and Lyft to other players that entered the fray catering exclusively to the air-taxi market. "Everyone's going home" you think to yourself.

Your rideshare zips along its route as you settle in for the ride home, with the skyscrapers of downtown receding in the distance and the lights of the city sparkling beneath you. You feel your smart wristband pulsing again. You look at your wrist and the glowing wristband projects "Answer" and "Decline" buttons hovering in the air. The Caller ID indicates that your spouse is calling you. You take off your wristband, lay it flat in front of you on the tray table in the seatback of your private pod and click the "Answer" air button. A holographic projection of your spouse appears – and even through the tiny hologram, you can tell that your spouse is delighted. "We just closed a jumbo umbrella policy" your spouse tells you excitedly. "Home, auto, boat, pet, disability, environmental hazard, and life insurance for almost everyone in ACME Small Business".

You know how hard your wife, an independent insurance professional, had been working to secure this new corporate client. Not only would it result in insuring multiple people at the same time, something that your spouse derived joy from due to the inherent nobility of providing a secure safety net for those insured, but also that the resulting commissions, in cash and cryptocurrency, would be a mini-windfall for your household. "That's terrific!" you respond jubilantly, before adding that this calls for a celebration. Your wife goes onto explain that their office AI assistant developed a holistic profile on every employee within ACME Small Business. Based on the AI profile, the algorithm developed a personalized bundled policy for each employee. As an example, for an employee who had recently announced that they were expecting their second child on social media, the AI assistant suggested that this individual was 78% more likely to purchase or reevaluate their life insurance, 63% more likely to move into a larger home in the exurbs from their current condo in the suburbs, 55% more likely to get a second dog, and 67% more likely to purchase a self-driving minivan. In the next decade, this employee's household, the AI predicted, was likely to purchase a second vacation home as well as a boat. In other words, there was an opportunity to continue doing additional business with individuals like this employee within the same company. The detailed and specific policy recommendations for each individual employee resulted in a better holistic insurance policy for the entire Small Business which could purchase business and commercial insurance, while providing group rates for every individual employee at a lower cost. It was a good day for your household.

Your wife asks if you have started dinner and suggests that a "fancier dinner" might be appropriate as a celebratory meal. "I agree with you" you say as you use your thumb to gently press the underside of your wedding ring. "Hey Alexa, start dinner, something fancy like a surf and turf, with a bottle of 'Moët et Chandon' please". You know for sure that you do not have nicer cuts of meat or lobster in the house, certainly not champagne to toast the occasion. However you're not concerned about needing to go shopping and then prepping for dinner. After all, your Amazon Prime Whole Foods delivery, with fresh lobster, filet mignon, and champagne will only take less than 10 minutes to reach your house via drone. "Alexa, cancel tonight's preprogrammed dinner" you remember to say. Surf and turf certainly does sound better than the *Chicken Kiev* that was programmed for preparation this Wednesday evening. With your connected refrigerator keeping track of refrigerated food and automatic delivery of fresh replenishments based on your usage and meal planning, visiting grocery stores had been a rare treat. "The convenience of life in 2046 is like no era before", you think to yourself. Your refrigerator kept track of your stock of refrigerated and frozen foods, your personal Amazon Astro (Brown, 2021) '44 robotic assistant (the 2044 model) performed or coordinated most of your household chores, including meal planning and keeping track of goods in the pantry, and you had a robotic lawn mower – albeit one that was ten years old - cut and treat your five acres of grass.

The younger generation questioned why they would need to step into a store when you can have anything delivered to your doorstep. However, you actually enjoy the occasional walk around a grocery store, taking your time and selecting your own produce. It is something that reminds you of growing up and doing the grocery store walkabout with your parents – pestering them for sugary cereal or candy as you walked around the aisles. "Dinner should be all set hon. Congratulations!" you tell your spouse who thanks you and tells you she needs to board her flight home. Wishing your wife safe travels, you end the call and settle in for the remainder of the 30 minute flight to the Northside Park and Ride. The sun sets over another Wednesday in the year 2046 as your Toyota air taxi flies you to your destination.

Your brief journey that offered you a glimpse of everyday life just twenty-five years into the future is coming to an end. Your journey of how you can prepare yourself for this everyday life is only just beginning.

SECTION TWO

UNDERSTAND THE NEED TO HAVE TO CHANGE

The Past is Prologue

Chapter Three – A Trip Through Time

It took the fictional character Dr. Emmett Brown, a custom-built "Flux Capacitor", fueled by plutonium that delivered 1.21 Gigawatts of electricity in order to power a DeLorean going precisely 88 MPH to enable time travel in the "*Back to the Future*" series. In the second installment of the movie trilogy, "*Back to the Future Part II*," Dr. Brown and Marty McFly travel to October 21, 2015 to a futuristic world as imagined by Hollywood in 1989. In the early 202x decade, some of their fantastical ideas envisioned in 1989 remain firmly entrenched in the genre of science fiction, yet there are many other advancements that even the most imaginative and creative minds could not have foreseen in 1989. As a matter of fact, some of the prognostications of future cities that were made in the film might well come to fruition in the next few decades. Due to technological advancements, science fiction continually becomes "science fact" today, and the "science fact" of today surpasses some of the wildest imaginations of science fiction from a few decades ago.

We took a trip to the future in the preceding chapters, without a DeLorean and a Flux Capacitor, just 25 years into the future to 2046. We covered a lot of ground and explored advances in technology that seemingly either sound too far-fetched, or too fantastical, or seem like the stuff of science fiction. But get this: **100% of the fantastical sounding technological advances we interspersed into our "the commute home on an ordinary Wednesday in 2046" story – each and every single technology referenced – is either available today, or is available in a pilot or trial version, or is under very serious active development at this very moment.** These technologies are being implemented - or at the very least - **exist in 2022** at a rudimentary level. The mythical "future" that we spoke of "yesterday", is "today", and in some instances, the "future" we spoke of "yesterday" *was* "yesterday". That "future" is already here. That "future" arrived and began setting roots in only the past 25 years and will continue its unprecedented and accelerating growth over the next 25 years, resulting in a world that seems almost fantastical in 25 years from now.

An Increasingly Unrecognizable World

"Any sufficiently advanced technology is indistinguishable from magic" (Clarke, 1962). British futurist and science fiction writer, Arthur C. Clarke called this thought-provoking statement his "third law" as a part of postulating his adages that have come to be known as "Clarke's three laws". This third law has certainly been on display throughout the 20th century, and only seems to be highlighted day in and day out in the 21st century. Consider the technological progresses achieved just thus far in the 21st century. The ubiquity of the internet and access to high-speed internet via a smartphone has, and continues to, transform our world. The affordability of these platforms have revolutionized how we work, how we live, and how we interact with one another. When we think about technology and technological advancement, it is easy to fixate on gadgets and gizmos. Consumer electronics are usually the gauge by which we measure the pace of digital change. It is the sheer ability to derive convenience and improve the quality of life with the use of technology that we equate to technological progress. We might expand our horizons to include our workplaces and the automation of the rote, manual work that we might have done in the past, and which is no longer needed due to digitization as a harbinger of our industry's place in the digital world. At times we end up missing the most common indicators of technological progress that have significantly improved our lives.

The advancements in medical care such as those where computerization and digital have provided quantum benefits in diagnostic, preventative, curative, surgical, and remedial procedures is something we often take for granted. Between these advancements in medical technology, always-available high speed internet, to the world at your fingertips with smartphones, wearable and connected devices, and all the "modern" conveniences that technology has provided to improve our quality of life, this is an unrecognizable world to people just one or two generations prior. Is there any doubt that our great-grandparents would be astounded to look upon the technological innovations available at our fingertips, or that just another generation before them would view the everyday technology we take for granted as nothing short of magic? Ask yourself whether your great-grandparents would even recognize today's world.

Our technological progress over the past 100 years has been nothing short of remarkable. It would be challenging to pinpoint any one technological innovation as the singular defining moment that propelled us into this accelerating pace of digitization in the 21st century. Some might say it was electricity, others might argue that it was the electrical grid, or the telephone, the automobile, air travel, or the internet. There is no right or wrong answer. Our technological progress has been driven by a series of catalysts that either built upon each other, or would not have been possible without each other. If you take a moment and think about these everyday technologies that we take for granted, and put yourself in the shoes of someone who lived 100 or 200 years ago, there will remain no modicum of doubt that Clarke's third law is a profoundly true statement; an ever-present "law" that itself continues to evolve. Consider how disoriented a person from 1810 or 1910 would be to discover the ability to video chat anyone across the planet at any time. Along the same lines, now consider how disoriented a person from 2010 would be to discover the ability to video chat anyone across the planet at any time - ***for free***.

Imagine the sheer delight of the individuals from 1810 or 1910 when the same device used for video conferencing plays music- any kind, any genre, any era of music at your beck and call. Think about how they would react when they watched their first streaming video on a smartphone? How would they react watching you order food using your phone and have it be delivered to your residence, or hail a ride-sharing service like Uber? In fact, while we are so fixated on the smartphone, consider that they would react with a mixture of amazement and bewilderment at "older" technologies (to us) like a lightbulb (and electricity in general), television, a telephone, a gramophone, a refrigerator, a microwave and cooking range, and yes, even an automobile, the so-called "horseless chariot". As a matter of fact, even in 1925, roughly only half of American households had electricity, with the other half still using gas light and candles to light their homes. How incredible is it that less than a hundred years after electricity was only available to less than half of the households in the United States, most of the world – including every major automobile manufacturer, following the success of Tesla Motors – is converting their entire automobile fleet to be fully or majority electric in the next decade? How would these generations react when they see an airplane flying overhead, watch footage of the moon landing, or learn of a human spacecraft – Voyager 2 – launched in August of 1977 and currently flying through interstellar space, or make sense of rovers driving around on Mars? We take technology and the pace of rapid digitization for granted today. Just think of the dramatic shift of our own attitudes towards technology over just the past twenty-five years, something we will explore further in this book.

No matter how close our "crystal ball" look into 2046 gets, I am almost certain that there will be inventions and innovations in the next twenty-five years that no one saw coming, just as no one saw the iPhone, smartphones in general, and always available high-speed internet coming in the year 1997 – only twenty-five years ago (as of this writing). To make sense of how far we have traveled in twenty-five years, most people reading this (as of 2022) sent their first-ever email, personal or business, less than twenty-five years ago. Reflect on that fact for a moment. The first email you probably ever sent or received was less than twenty-five short years ago. Think about the number of emails you might have sent and received since that first one. Do you think you have sent and received thousands of emails? Tens or even hundreds of thousands maybe? Do you recall the novelty you felt when you authored your first-ever email. It is unlikely that most people would even remember who their first email was addressed to, and who their first email sender was.

The popularity of text messaging – especially in the United States – is even more recent than email. Although SMS and MMS text messaging had existed since the late 1990s, it did not become widely used and wildly popular until the middle of the first decade of the 2000s. The arrival of the iPhone in 2007, ushering in a new age of smartphones, made text messaging incredibly commonplace. And of course, as expected, mobile phones soon rendered landlines obsolete. Voice-Over Internet Protocol (VOIP) phones, which most digital phones across corporations are today – further challenged an already-challenged landline landscape. Now think about the "Generation Z", those born between 1997 and 2012. This is a generation that has had smartphones since Elementary School. Think about how this generation uses their smartphones – almost as an extension of their own cognitive selves. This generation has never used a calling card, much less a payphone. They do not comprehend the concept of phone booths (or "Clark Kent" having to change into "Superman" in a phone booth). They have never had to self-regulate when they call their friends because the concept of "airtime" and "minutes" becoming cheaper after 9 PM is alien to them. They only know calls, video and audio, and text messaging to be always available and always "free". As of this writing (2021), this generation is not even legal drinking age in the United States of America, and just barely became eligible to vote in the 2020 US Presidential Election.

From Science Fiction To Science Fact

We took a trip into the future, just twenty-five years into the future in the first two chapters of this book. To understand where we are headed, it would be helpful to reacquaint ourselves with where we currently are. To be able to appreciate where we currently are, it is in turn helpful to catch our breath and take stock of how we got here today. In order to do that, let us take another trip. This one does not need to be confined by the bounds of our imagination, and you do not need to envision things that once seemed too fantastical to be reality. All we need to do is just travel back in time about fifty years and work our way forwards, focusing specifically on just the past twenty-five. I assure you that when we travel back to the present from twenty-five years ago, you will fully comprehend and have a better sense of appreciation for how, technologically speaking, the next twenty-five years might play out.

Daily life for a large portion of the world's population as we enter the third decade of the 21st century revolves around technology. Technology has become the one common thing that unifies people across the world, the only variation being in the amount that people are reliant (or downright dependent) on technology in their daily lives. Although there is absolutely a "digital divide" in the world today with the level of access that individuals across the socio-economic spectrum have to affordable technology, it has by-and-large permeated the wealthiest and the poorest nations. This access to affordable technology, including access to high-speed, always available, unrestricted/uncensored (in most countries) internet, continues to expand at an accelerated pace. The evidence for this bears out in just the five year period between 2015 and 2020, a time <u>before</u> a once in a hundred year worldwide COVID-19 pandemic forced the world to digitize at scale – from work to school – nearly overnight. The following are some quick comparative figures illustrating the contrast between 2015 and 2020, according to research by Hootsuite and 'We Are Social' 2020 (Hootsuite, 2020):

a. World Population: 7.2 billion in 2015 vs 7.77 billion in 2020
b. Number of Internet Users: 3.01 billion in 2015 vs 4.57 billion in 2020
c. Number of Mobile Phone Users: 3.6 billion in 2015 vs 5.16 billion in 2020

What jumps out at us from these statistics is that for the first time in our history, as of 2020, more than half the population of this planet had access to high-speed internet and a greater number were mobile phone users (about two-thirds of people on Earth). The research (Hootsuite, 2020) also revealed that an average user spends slightly less than 7 hours online per day, and social media, in aggregate, boasts an active user base of 3.81 billion people (slightly less than half of the world's population). We can only begin to fathom what these internet coverage statistics might look like in the next decade.

Disrupting the Digital Disruption

With incalculable and tragic loss of life and with the financial calamity that it wrought onto millions of people across the world, the horrific COVID-19 pandemic will be remembered a seminal moment in our history. Starting in early 2020 and having stretched into 2022 (as of this writing), the pandemic disrupted the digital disruption that was already underway. No one can imagine having had to have dealt with this pandemic twenty years ago. Twenty years ago, hundreds of millions of people around the world would have been unable to work from home. Twenty years ago, hundreds of millions of children and college students across the world would have been unable to receive education via the internet. The pandemic's severity and the subsequent global lockdowns, already significantly impactful to the world, would have been felt tenfold had we not have had the internet, or if the internet had not been robust enough to support millions of more people online and concurrently online.

From a digitization perspective, this pandemic undeniably impacted every single person working in every single profession, in every single company across every single sector of every single industry. The digital transformation of these industries, especially those who had historically been laggards in adoption of digitization, was "hyper-accelerated" by this pandemic. Some sectors of the industry that had been on "cruise control" have had to scramble to reprioritize their digital transformation journeys in order to ensure they stay relevant and topical. From an "Industrial Revolution" perspective, the COVID-19 pandemic has served as a "Micro Industrial Revolution" (in the grand scheme of things) smack dab between the end of the "Third Industrial Revolution" and the start of the "Fourth Industrial Revolution" (as explained in the following chapter).

Connected Everything

Our global reliance on being a connected world, being tethered to each other and have access to unlimited information at our fingertips via the internet has been irrefutably proven through this pandemic. For right, wrong, or indifferent, the worldwide web is the backbone of economies, societies, and cultures across the world today. We use it for shopping, for commerce, communication, entertainment, business, education, connection, collaboration, and companionship. It is an inexorable part of us today and the next evolution of this connected world is that these connections will find themselves into every single facet of our lives. From Smart TVs and Smart Refrigerators to personal digital assistants such as Amazon's Alexa -and imminently the Amazon Astro personal robot (Amazon, Amazon Astro, 2021) - and Google Home, there are a lot of things on our person and in our homes that are connected to the internet today, and this will become the norm for all "things" as we forge forward. These always-on, always-connected "things" that continually transmit and receive data from and to the internet are what gave rise to the term "Internet of Things" ("IoT" for short). Now consider that this thing called the internet, something we are all so connected to and *connecting* through was born less than 50 years ago. "Prior to the Internet being a place for posting, tweeting and streaming it was known as the US Defense Department's Advanced Research Projects Agency Network (ARPANET), or the 'Eve' network, which was first published in 1967. And two years later, it become a way of interconnection for four university computers. ARPANET's initial purpose was to communicate with and share computer resources with mostly scientific individuals at the connected institutions – this included MIT, Harvard and University of California, Los Angeles." (Liberatore, 2016).

This is what the entirety of the internet looked like in May of 1973. Image credit ARPANET 1973 (Public Domain).

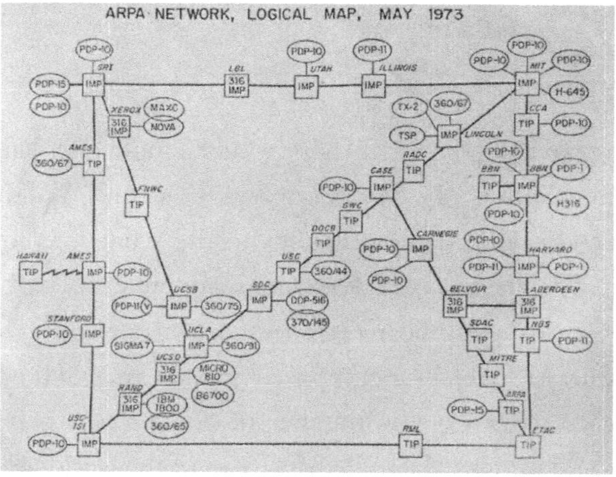

Image 1 - ARPANET (1973)

Superimposed onto a map of the United States, these connection points that formed the entirety of the internet in 1973 looked as depicted in Figure 2 below.

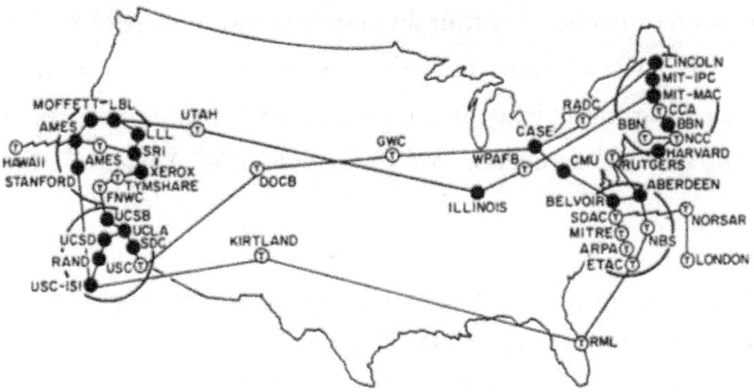

Image 2 - ARPA / Image Credit: ARPAnet Completion Report

A similar map of the internet today would simply look like a "beautiful mess" of connections. It has been said that the World Wide Web connections today look quite analogous to the synaptic connections of a human brain. The internet is so deeply meshed into every part of the planet that if we were to superimpose these connections on top of a world map, like we had with ARPANet superimposed onto the map of the United States in 1973 (Image 2 above), the number of connections between internet endpoints would obscure the underlying map of the world (and connecting into space due to the number of satellites transmitting data to and from Earth as well). A simple search on Google Images will yield a bounty of images depicting everything from internet connection endpoints, undersea cabling, internet traffic routes, and most and least active areas across the world, etc. Using such an image in this book, in a black-and-white print would blot out the sheer jaw-dropping allure that the internet is today, akin to looking at a renaissance masterpiece in black and white. The internet today is controlled chaos in all its beauty.

While the internet has certainly been transformative for society, culture, and quite certainly having fundamentally thrust our species into new and unchartered waters, technological progress over the late 20th century has been continual and exponentially accelerating. In the following chapter, we will examine some of the basic and often unheralded technological advancements over the latter half of the 20th century. The chapter shall then focus in on some everyday products that we have become used to leveraging on a daily basis – products that did not exist twenty years ago. Once again, it begs the question, imagine if you were transplanted into today's world from a hundred years ago, a fifty years ago, or even twenty years ago – would you recognize all of the technology that we take for granted today? Now imagine being transplanted twenty years, fifty years, and a hundred years into the future – is there little question that this future world would likely be as unrecognizable to us as today might several decades ago?

Chapter Four – Today is Yesterday's Tomorrow

"Once a new technology rolls over you, if you're not part of the steamroller, you're part of the road." - Stewart Brand, American author and founder of several organizations.

Technological advancements over through the 20th century have truly been remarkable. We will revisit this fact later in the book but consider that it took only 66 years between the Wright Brothers inventing flight in 1903 to humans setting foot on the moon and returning to Earth safely in 1969. The Apollo 11 mission that carried the first humans to the moon used a sophisticated computer system. Known as the Apollo Guidance Computer (AGC), this computer system was considered to be quite a cutting edge computer at the time. As of 2022, slight over fifty years later, the Apple iPhone has over a million times more temporary memory than the AGC, can store over 7 million times more data in storage, and is said to be well over 100,000 times faster (Kendall, 2019). Just as is the case with the internet, computing power and storage has become commonplace, exponentially faster, and order of magnitude more affordable to the masses. Just as with the computational power that resides within the iPhone, or any smartphone today, these technologies have also become incredibly smaller – to the nth fractional percentile of their scale just a few decades ago. The image below, courtesy IBM and CUNY (Cheng, 2017) serves as evidence of this phenomenon.

Image 3 - IBM Storage Device (5 MB, 1956)

The above image (Image 3) is of an IBM storage device in 1956. Boasting a massive (for the time) 5 megabytes (MB) storage capacity, the IBM 350 disk storage unit required several hefty gentlemen, muscles, a ramp, and a truck to transport from one location to the other. You would be challenged to find a single digital photograph in your smartphone's Camera Roll (iPhone)/Photo Gallery (Android) that is less than a 5 MB file size. The handful that you might find if you are lucky are likely older photos taken using an older smartphone and carried forward using the Cloud, or those of much lower quality (pixels). 5 MB of storage was considered significant in the late 1950s. In contrast, fast forward to the 1990s through the early 2000s. The age of personal computing and the need for businesses and consumers to create, store, and exchange digital files such as word processor documents, spreadsheets, and slideshows led to the development and incorporation of floppy disks as a form of external, removable storage. Floppy disks that stored 1.44 MB of space eventually gave way to CDs that allowed for several hundred megabytes of removable/portable storage.

Image 4 – Floppy Disks (late 1990s/early 2000s). Image credit: George Chernilevsky, Public domain, via Wikimedia Commons

From a humble 1.44 MB in 1998, the Micron® company, in 2019, released their c200 1TB microSDXC UHS-I card. The world's highest-capacity microSD card, the Micron® product delivers one terabyte (1TB) of high-performance removable storage.

Image 5 – Micron® 1TB microSD card. Image credit: Micron® Technology

The image, Image 5 above, courtesy Micron®, depicts a microSD storage card with a whopping 1 TB of storage. Just how much more storage capacity is the IBM 5 MB disk storage unit from 1956 to the 2019 Micron® microSD storage device? 1 TB equals 1,000,000 MB. This equates to 200,000 5 MB drives. For perspective, in 1956, you would have required an IBM hard drive roughly the size of a 40-story building to equal the storage capacity of the Micron® microSD card, which is roughly the size of a fingernail.

The Past Twenty-Ish Years And Less Than The Past Twenty Years

We do not need to look too far into the past to reacquaint ourselves with this surreal era of technological change. Just the past twenty years, give or take a couple of years, are fairly indicative of this accelerated digital journey. From that perspective, here are a few things that are twenty-ish years ago, and some that are less than twenty years old (as of 2022). It is mind-boggling to take a moment and reflect on that fact that just the past twenty years, give or take a couple, have provided us with everyday technological products and services that we take for granted today. It is astonishing to think about the fact that the products and services depicted below did not exist at the start of the 21st century.

Slightly Over 20 Years Old

i. The USB: Continuing the conversation of external removable and portable storage devices (known as removable media), USB devices are ubiquitous today. Universal Serial Bus (USB) is an industry standard that establishes specifications for cables, connectors and protocols for connection, communication and power supply (interfacing) between computers, peripherals and other computers (Wikipedia, USB, n.d.). The USB flash drive became extremely popular as an external media device in the early 2000s, replacing floppy disks and CD-ROMs. This technology did not exist before 2000 and was patented by Israeli-based M-Systems in 1999/2000 (CIE, 1999). It is indisputable that M-Systems' USB flash drive, as displayed in Image 6, revolutionized portable data storage with a storage capacity of 8 MB, five times the memory of most floppy disks at the time.

The USB flash drive, a common removable, external media device for the past fifteen years at least, before Cloud storage emerged as a viable, evolutionary alternative, and has served to continually reduce reliance on external media, is less than twenty-five years old.

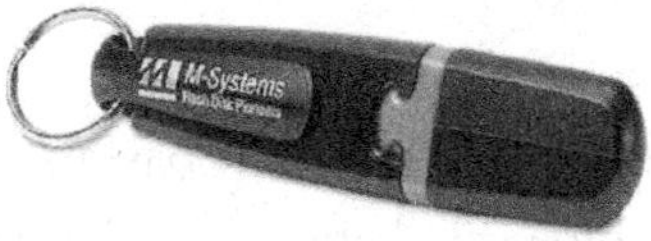

Image 6- M-Systems USB Flash Drive (2000). Image Credit: Center for Israel Education (CIE)

ii. Bluetooth: From your Apple AirPods or any wireless headset, to connecting your smartphone to your vehicle for streaming music, phone, and GPS – can anyone imagine the difficulties of tethering to a myriad of our technologies together today without Bluetooth being available to us? Bluetooth is a wireless technology that allows the exchange of data between different devices within a short amount of distance. Most Bluetooth devices have a maximum range of about 30 feet, a distance that is reduced when there are obstacles to the wireless signal present. Bluetooth devices are said to be generally safe against hacking because they operate on any of various different frequencies and hop" between these frequency bands several hundreds of times per second. Bluetooth, created by Jaap Haartsen and developed by the Bluetooth Special Interest Group (Bluetooth-SIG, n.d.) was introduced on May 7th 1998. In just twenty-three years, Bluetooth has gone from an emerging technology standard, complementary to Wi-Fi, to being an integral part of nearly every computing device, from PCs to smartphones.

iii. Mobile Phone with Camera: How amazing is it to think about a time where mobile phones had no cameras? For anyone over the age of thirty, it is quite common to remember having to purchase film, take photos on a camera that you had to lug around, drop the film off for development, and wait 3 days to a week before you were able to see the photos you clicked. Traveling for vacations, or special occasions meant that you had to carry your camera with you – and make sure you had adequate film. Should your photos not be to your satisfaction when you received them after having spent time and treasure in getting them developed and printed at a store, you had absolutely no recourse. There was no opportunity for a redo since there was no way to preview the photos, delete ones that do not pass muster, and click new ones. Most over thirty years old also remember passing a scenery, an unexpected sight, or a sunset, and being unable to capture and share the moment (sorry to borrow your catchphrase, Kodak) since there was no camera handy.

The age and pervasiveness of social media also served as the catalyst for our photography boom. With eager social media users seeking to share their experiences – from their vacations, food, and baby pictures with friends and family – entire social media platforms such as Instagram and Snapchat are based around photos (and videos) as core content. But here's the kicker. While Apple and the iPhone might have revolutionized the camera/smartphone combination, the first mobile phone with a camera embedded in it was the Sharp Electronics J-SH04 J-Phone introduced in November of 2000 (or, according to some, the Samsung SCH-V200 that was introduced 5 months prior). "Many consider the J-Phone to be the first true camera phone to come to market because, in addition to containing a fully integrated phone and camera system, the J-Phone allowed you to share pictures directly from the phone, with no computer needed." (Weitz, 2017). The first "true" camera phone is less than 22 years old as of 2022. The image below, courtesy B&H Photo and Video, depicts the Sharp J-SH04 J-Phone.

Image 7 – Sharp J-SH04 J-Phone. Image Credit: B&H Photo and Video (Weitz, 2017)

While we are on the topic of smartphones with cameras, it is worthwhile to note that it was not too long ago where humankind crossed an important milestone. It was only about a decade ago that statistically speaking, thanks to these camera-equipped smartphones, we as a civilization took more photographs in a single day than all the photographs taken during non-digital means since the invention of the camera. According to Rise Above Research, there would have been an estimated 1.4 trillion photos taken in 2021, projected to hit 1.61 trillion in 2023 (Lee, 2021). Nearly 91% of these photos will have been taken via a smartphone, 3% using tablets, and 6% using a digital camera (Pantic, 2021). To put that number of a trillion in perspective (photography pun not intended), if you continually took one photo per second, it would take you approximately 12 days of continuous photo taking to take 1 million photos. If you continue to get to a billion photos, it would take you approximately 32 years of photo taking. To get to a trillion photos, taken continually at the rate of one photo per second, it would take you approximately 32,000 years. In addition to the actual clicking of photos, there is also the question of storage and saving these memories. There have been nearly 7.9 trillion photos stored on computer drives, PCs, mobile devices, in the cloud, printed out, etc. in 2021. Assuming you stored one photo continually per second, it would take you approximately 250,000 years to reach that number. This number is expected to grow to approximately 10 trillion photos in 2023. As a time frame of reference,

250,000 years ago, the greatest technological advancement on this planet was that Neanderthals were mastering fire making (Pantic, 2021).

Photos and photography are a good gauge of our technological progress over the past two decades because they are so relatable. We likely do not go a single day or a full week without taking, sending, receiving, storing digital photos, or posting them on social media. We have certainly come a long way since the first photograph was taken in 1826. As depicted in Image 8 below, the world's first photograph was made in a camera was taken in 1826 by Joseph Nicéphore Niépce. The photograph was taken from the upstairs windows of Niépce's estate in the Burgundy region of France (ARCHAMBAULT, 2021).

Image 8 – The World's First Photograph (1826). Photo Credit: PetaPixel (ARCHAMBAULT, 2021)

Continuing the photography theme through history, Image 9 below depicts the first photograph of a human. Capturing the image of a person getting their shoe shined, this photo was in taken by Louis Daguerre in 1838, capturing the Boulevard du Temple, a thoroughfare in Paris, France (Krulwich, 2010).

Image 9 – The World's First Photograph Featuring a Person (1838). Photo Credit: PetaPixel (ARCHAMBAULT, 2021)

Now where does this journey leave us? In less than 200 years since the first photographs were taken, Image 10 below, of a young person seemingly regurgitating a sparkly rainbow, for perspective, is one of the most popular Snapchat filters of all time.

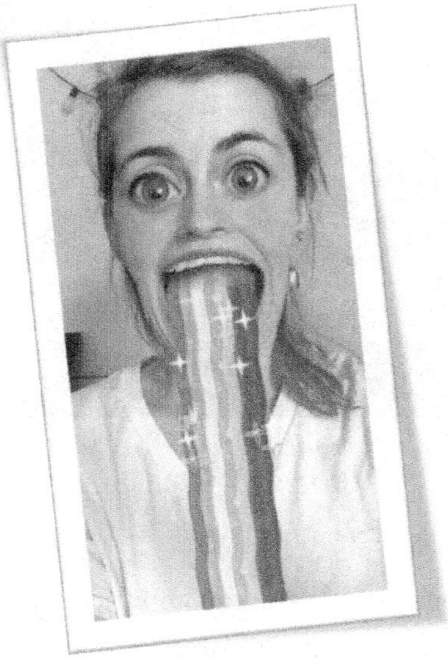

Image 10 – One of the most popular Snapchat filters in history (2018)

Less Than 20 Years Old

i. Already (nearly) passé – Blu-ray Disc: The Blu-ray disc superseded DVD-Video. DVDs were the dominant home video format across the world in the early 2000s until it was supplanted by Blu-ray discs. Coinciding with the rise of Blu-ray, DVD sales plummeted more than 86% from 2008 through 2019 (Witten, 2019). Blu-ray is capable of storing several hours of high-definition video (HDTV 720p and 1080p). After winning the high-definition video disc wars by beating out the competing HD-DVD format, Blu-ray discs and players became a commonplace consumer electronics item – featured in video game consoles such as the Sony PlayStation, in PCs, and as stand-alone video players.

For a platform that has dominated for about a decade, Blu-ray itself is headed towards obscurity, having been displaced by streaming high-definition video services. From Netflix to Hulu to Amazon Prime to HBO Max to Peacock, and the ability to download video games directly to the console or play online – the streaming space itself is getting increasingly saturated, threatening to completely crowd out Blu-ray. After all, who wants to maintain additional hardware and worry about storing, transporting, and caring for physical media? Just like email displaced physical letter writing, market demand for physical media, be it for business or entertainment, is waning rapidly. Having released the first commercial Blu-ray video disc player in the late 2000s, Samsung announced in 2019 that they would not be producing any new Blu-ray players for the US market. The streaming media market, just in the United States alone, is projected to bring in $42 billion in revenue in 2025. Netflix dominates the US market as of the end of 2021, with 75 million subscribers, followed closed by Amazon Prime Video (50 million subscribers), Disney+ (45 million subscribers), Apple TV+ (40 million subscribers), and Hulu (35 million subscribers) as the top 5. Globally, YouTube is the most popular video streaming service in the world with two billion active users, followed by Netflix with 195 subscribers worldwide (Curry, 2022).

ii. The iPhone: Steve Jobs announced the iPhone to the world on January 9th 2007. Since then, the iPhone, widely considered as a catalyst for the mobile computing, the app economy, the world's dependence on smartphones, and always-available internet, has sold well over 2.5 billion iPhones (projected number since Apple stopped reporting iPhone sales figures in 2018). Although touch screen devices existed quite broadly long before the iPhone, the iPhone was the primary driver to popularize touch screen devices for mobile phones. As we will discuss later in this book, the popularization of this touch and tactile technology served as the beginning of the end for one of the most popular corporate-use mobile devices at the time, the BlackBerry, as well as the most successful mobile phone brand on the planet at the time, Nokia. "The first-generation iPhone was described as 'revolutionary' and a 'game-changer' for the mobile phone industry and subsequent models have also garnered praise. The iPhone has been credited with popularizing the smartphone and slate form factor, and with creating a large market for smartphone apps, or 'app economy'. As of January 2017, Apple's App Store contained more than 2.2 million applications for the iPhone" (Wikipedia, iPhone, 2022). With its ecosystem via the App Store, iTunes Music, and portable iOS operating system, the iPhone revolutionized the smartphone industry, and drove Apple to attain heights that were once considered impossible in the late 1990s as the company struggled with being on the verge of bankruptcy. The iPhone, as impactful as it has been for our world – personally and professionally – is less than twenty years old.

iii. Today's Tech Giants: Towards the end of 2021, five of the six biggest companies in the world in terms of market cap, with the sole exception of Saudi Aramco, were technology firms. Each of these five technology companies were worth more than a trillion dollars as of December 2021. If you expand the list to the top companies in the world, with the exception of the aforementioned Saudi Aramco and Berkshire Hathaway, are all technology companies. With 80% of the world's top companies being technology firms, is there any modicum of doubt about the power of technology in the 21st century? Just in the year 2021 alone, Apple, Microsoft, Alphabet, Amazon, Tesla and Facebook added a combined $2.9 trillion to their collective market caps (Vega, 2021). Consider some of the most common technology companies that we as consumers – both personally and professionally – interact with and rely upon today. In this new technology powered economy, Netflix established itself as the world's largest movie house without having owned any movie studios. Facebook established itself as the world's largest content provider, but owns no content. Uber and Lyft established themselves as worldwide transportation systems, displacing and competing with taxi cabs, but own no vehicles. UberEats, GrubHub, and other food delivery companies own no vehicles. Airbnb, vrbo, and other such companies provide intense competition of hotels, motels, and resorts, but own no properties of their own. WhatsApp (owned by Facebook's parent, Meta) is the world's largest telephony platform, but owns no phone lines. Tesla is to the electric vehicle market what the iPhone is to the smartphone market. Apple and Amazon are two of the most recognizable brands across the world. In fact, Amazon, a technology company, is basically at competitive war with established firms across nearly every industry. Now here's the kicker. With the exception of Apple (founded in 1976), Microsoft (founded in 1975), Amazon (founded in 1994), and Google (founded in 1998), none of the companies that we rely on today - the world's largest and most recognizable companies – even existed twenty years ago. Twenty five years ago, Google did not exist, Apple was teetering on the edge of bankruptcy (before Bill Gates – Microsoft – saved the company with a $150 million investment in 1997), and Amazon was still primarily an online bookstore.

As disorienting as it is to think about a life that existed before these technology players became an inexorable and intimate part of our lives - whether it is a Tesla, a technology company that manufactures vehicles, or an Apple that had to pivot from "cool" Macintosh computers to spearhead different forms of computing in the consumer electronics sector - each of these companies have a few things in common. First and foremost, it is indisputable that each of these firms are all intensely customer focused. They not only know their customer, but it is safe to state that they understand their customer. Whether it is aided by data and analytics – such as by developing an accurate customer profile based on what posts get most reaction and engagement such as with Meta/Facebook, or by tracking purchasing patterns such as with Amazon – these companies take the time and effort to understand their customers. The second commonality across these companies that traverse disparate sectors is that each of these companies either embraced disruption (as is the case with Apple entering consumer electronics and Netflix when they switched to being a streaming service), or drove disruption within their industries (Uber disrupting taxi cabs, Airbnb disrupting the hotel industry, Tesla disrupting automobiles) or across other industries (Tesla developing SpaceX and disrupting governmental space agencies, Amazon purchasing Whole Foods and causing a seismic shift in the grocery business).

The Future Was Yesterday

To predict what a journey into the future might look one, one needs to only take a peek into the past. Even the very recent past is a good indicator into what the future might hold in store for us. So why should we care about this technology-fueled future? We seem to have done fine thus far over the past twenty five years – technology is an intimate part of our lives, at home and at work. We have seemed to incorporate digitization into work and adopted as technology progressed all around us. The simple answer is because this pace of digitization is accelerating and accelerating at a pace that we might, without some foresight and forethought, be unable to keep up with. Imagine what happened to every employee at Blockbuster – from the CEO and the executive team, all the way down to the high school student employed as a cashier at his or her local Blockbuster store.

Netflix usurped the entirety of their business and left them in pursuit of a different job. I am sure that some of those employed at Blockbuster were able to seamlessly segue their corporate skills to a different industry and sector of the economy. And yet, I am also certain that there is are a measurable number of former Blockbuster employees that faced tremendous career disruption. Potentially having to scramble to retrain themselves, go back to school for new skills, start at a lower position in a different industry, or accept lower wage employment in another sector – I am certain that life during this disruption might have been energizing and exciting for a few, but abjectly miserable for a lot of them. The stress of witnessing your firm and industry (video rental) be rapidly vanquished by an upstart and by a product that few truly understood at the time, and having to find employment to support themselves and their families could not have been easy for any of those employees. It might have been especially stressful for those generations bookending the Blockbuster workforce. For those in their 50s and 60s, having to reskill in a compressed period of time with an uncertain future and uncertain direction, while having to deal with a mortgage, healthcare expenses, childcare expenses, college tuition, etc. might have been quite onerous. However, for those early in their careers, in their early 20s or so, where working at Blockbuster corporate might have been their first job out of college, finding another job with little experience and the "taint" of a failed sector as their sole experience might have also been quite challenging.

The period of Blockbuster's collapse was also around 2010, when the US and world economy was just barely recovering from the awful "Great Recession" when the unemployment rate in the United States according to the US Bureau of Labor Statistics (BLS) was at a staggering 9.3% with 14 million people unemployed (BLS, 2011). Can you imagine having to desperately look for a job and perhaps lack the skills employers were looking for during that period in such a labor market? I used the Blockbuster illustrative example not to serve as a "downer" and bum anyone out, but to depict the driving force behind this book and the urgency with which we – each of us – regardless of sector or industry – need to be proactive and vigilant about owning our careers in the 21st century. The Blockbuster story has been played out over the past twenty years across companies, sectors, and industries. No industry is technology disruption proof and certainly no career or job in any industry is technology disruption proof.

Even without access to a DeLorean fitted with a Flux Capacitor, a trip to the future can be at least somewhat figuratively undertaken thanks to the past. The stuff of science fiction fifty years ago is science fact today. The stuff of science fiction today will become science fact in a fraction of that in the next twenty five years. While this breakneck pace of change has brought about a more connected world – with all the positives and negatives, this pace will continue to increase, with the potential to dramatically alter what the world looks like in the next twenty-five years. It is said that this "all-technology, all-digital" era we have all just entered is called the "Fourth Industrial Age", a concept that we will define and explore further in this book. And we, as a workforce that is comprised of several cohabiting generations that are transitioning from the "Third Industrial Age" into the "Fourth Industrial Age", are utterly underprepared for it.

Chapter Five – Welcome To The Fourth Industrial Revolution

We ended the preceding chapter stating that we, as a society, are utterly unprepared for the shape of things to come. Our being ill-equipped to navigate through the end of one "Industrial Revolution" and the start of a new one "Industrial Revolution" is not unique to our generation. As we will explore in this chapter, with the three preceding "Industrial Revolutions" all through to the "Fourth Industrial Revolution" that we are entering, there have always been millions of workers that find themselves to be a workforce in transition. Subject to the external forces of change around them as one "Industrial Age" shifts into the next, these workers are ones that experience the most change and disruption as compared to those who find themselves in the *middle* of any particular "Industrial Revolution". It is my anecdotal opinion that the ten years preceding the officially recognized end of one "Industrial Revolution" and the start of the next one, and ten years into the new "Industrial Revolution" are two decades that are marked by significant disruption. These twenty years are deterministic to how an entire workforce is going to prove to be a viable one starting with the second decade into the new "Industrial Revolution".

Workers who fail to learn new skills, augment their core skillset, or embrace the disruption and reinvent themselves are the ones who will be in the most precarious and perilous positions. Their careers are the most susceptible to being disrupted or disintermediated, either by technology and automation, or by other workers with more current skills that the new "Industrial Revolution" demands. And while the transitionary nature of our workforce is not unique, our workforce is in a unique situation by entering a unique "Industrial Revolution".

The "Fourth Industrial Revolution" will bring with it a 'crushing cascade of career changes'. These changes have the potential to prove quite positive and profitable for some careers and could be crippling to others. The basic difference between those careers who will make it and who might not, more specifically, the people in these careers who will make it versus those who might not, comes down to understanding, planning, and preparation. There can be no action without preparation, and plans can be ill-conceived without a holistic understanding of what it is that we, as workers, are truly going to be imminently dealing with. Our ability as workers - regardless of vocation, company, sector, or industry – to fully grasp the upcoming disruptions, develop our own plans to prepare, and instill in ourselves the capabilities to adopt and adapt rapidly to these changes – are what will help us to minimize risks to our careers and maximize rewards. To be able to do that, we need to understand what the "Fourth Industrial Revolution" means, and gain an appreciation of the three industrial revolutions that have come before.

What Is An Industrial Revolution?

An "Industrial Revolution" is characterized as an era where advancements in how entire industries operate are upended by advancements in how industries manufacture and/or distribute their products and services. The era is termed as a "revolution" simply because the introduction of new tools, technologies, or processes are considered to bring about a change dramatic enough that it has no precedent and these changes typically spawn multiple other industries as a result.

These macro changes also have a tendency to disrupt entire industries, economies, societies, cultures, and trigger shifts in the balance of global socio-economic powers. According to Britannica, "Industrial Revolution, in modern history, is the process of change from an agrarian (agriculture and farming-based) and handicraft economy to one dominated by industry and machine manufacturing. These technological changes introduced novel ways of working and living and fundamentally transformed society." (Britannica T. E., 2021). To get a handle on the "Fourth Industrial Revolution" that we are only just entering ("just entering" as measured against the broader timeline on the average length of each Industrial Revolution), it is important to understand how different this era is from the three that came before it. Note that there is some debate on precisely when the "First Industrial Revolution" commenced, and even more debate on when the "Fourth Industrial Revolution" started (the difference in the start year here is too minor to be considered consequential – some state the commencement as 2016, others as 2020).

There is also a school of thought that seeks to categorize the era that a majority considers to be the "Fourth Industrial Revolution" as the "Fifth Industrial Revolution". In this definition as stated by a small minority of thought leaders, they postulate that the era from 2000 through 2020 can be considered as the "Fourth Industrial Revolution" – the rise of affordable, high-speed, always-on internet and the subsequent ubiquitous nature of this internet at the fingertips of most across the world via smartphones. For our purposes, we will align with *vox populi*, and in keeping with thought leadership of a majority, look at four - not five - "Industrial Revolutions".

The Industrial Revolutions

1. The First Industrial Revolution (Industry 1.0):

The "First Industrial Revolution" commenced in the late 18th century and continued through the early 19th century. Starting sometime between 1760 and 1784 (lack of consensus on exact start year), "Industry 1.0" was the era of Mechanization. Industrialization/Mechanization replaced entire agriculture-based economies and societies transformed with this change. Steam power, an undeniable revolution at the time, gave rise to the development of factories engaged in mass manufacturing of products and services. Steam power made rail travel eventually possible, and rail travel allowed for commerce and people to traverse large distances in a relatively shorter span of time. With people (eventually) being able to travel (and resettle), migration led to the development of industrial cities.

Some New England cities in the United States were settled based on their proximity to rivers and ability to provide access to crisscross the Eastern Seaboard for multiple industries, from logging to textile development. Loom weaving, and specifically the manufacture of textiles, took root across these "connected" areas. The use of steam power and mechanization of production allowed for the achievement of 8 times the volume as compared to what used to generate textile threads on rudimentary spinning wheels. Citing an article from Desoutter Industrial Tools, "The use of steam for industrial purposes was the greatest breakthrough for increasing human productivity. Instead of weaving looms powered by muscle, steam-engines could be used for power. Developments such as the steamship or (some 100 years later) the steam-powered locomotive brought about further massive changes because humans and goods could move great distances in fewer hours." (Desoutter, n.d.)

2. The Second Industrial Revolution (Industry 2.0)

The "Second Industrial Revolution", or "Industry 2.0", started around the late 19th century and continued through the middle of the 20th century. This was the era of mass production and the assembly line. Electricity and petroleum gained prominence - and eventually, dominance – as sources of energy. Amongst the greatest innovations introduced about midway through this epoch, that is said to have lasted from 1870 to 1969, was undeniably electricity. The generation, distribution, and conversion of entire societies to be an electric-powered population is one of the marvels of the 20th century. It will be almost impossible to find even a single thing that we and our lives rely on today – from our personal lives to our professional lives – that does not leverage electricity. Electricity is arguably the backbone of the world economy and peoples' lives today and it is an inexorable part of society. This era also witnessed the explosive growth of modes of transportation. These petroleum based vehicles, from automobiles, to commercial trucking, to airlines fundamentally transformed how people live, work, and engage in leisure activities. The world metaphorically shrank more in this era than any period in history before it. Household appliances became commonplace.

The "Second Industrial Revolution" introduced an unprecedented slew of technological advancements and changes that until then (and until the "Fourth Industrial Revolution"), the world had never experienced before. Cities grew at an exponential rate, new industries emerged, factories sprawled, the concept of consumer electronics took hold, and in a reflection of industrialization as compared to agriculture-based economies, and people's lives were said to be "regulated by a clock rather than the sun" (Desoutter, n.d.). Excerpts from an article offered by The History Channel summarize this era well - "Rapid advances in the creation of steel, chemicals and electricity helped fuel production, including mass-produced consumer goods and weapons. It became far easier to get around on trains, automobiles and bicycles. At the same time, ideas and news spread via newspapers, the radio and telegraph. Life got a whole lot faster. Factories produced sewing machines for home use, steel girders for skyscrapers and railroad tracks that cut through the plains and mountains. Long-distance transportation networks connected by rail, steamship and canals opened new markets for farmers, factory owners and bankers who could bring America's natural resources to a global marketplace. For the first time, goods from the American heartland could be shipped long distances, eliminating the need for local bartering systems. Household items like soap, butter and clothing that used to be made at home started being made in factories as well. At the same time, all kinds of goods became standardized for the first time. (Niiler, 2019)"

The most famous example of mass production and the assembly line is the Ford Motor Company and the Model-T Ford. "Henry Ford (1863-1947) took the idea of mass production from a slaughterhouse in Chicago: The pigs hung from conveyor belts and each butcher performed only a part of the task of butchering the animal. Henry Ford carried over these principles into automobile production and drastically altered it in the process. While before one station assembled an entire automobile, now the vehicles were produced in partial steps on the conveyor belt - significantly faster and at lower cost." (Desoutter, n.d.). According to excerpts from a course about the "Second Industrial Revolution" on the website study.com, "While the First Industrial Revolution caused the growth of industries, such as coal, iron, railroads and textiles, the Second Industrial Revolution witnessed the expansion of electricity, petroleum and steel. Many of the changes that occurred during this period had to do with new products simply replacing old ones. For instance, during this time, steel began to replace iron. Steel was being utilized for construction projects, industrial machines, railroads, ships and many other items. Steel production made it possible for rail lines to be built at competitive costs, which further spread transportation.

It is hard to imagine a time when electricity was not a common luxury. However, the beginning of the Second Industrial Revolution was just such a time. Before the introduction of public electricity, candles and gas lamps were used to light homes and factories. Many activities were simply done in the daytime. The use of electricity fundamentally changed the way people worked and lived. The first efficient commercial electrical generators were used in the 1870s. In 1881, Great Britain was the very first country to install a public power station. Beginning in 1910, it was possible to power a residential area from one single station. The introduction of public electricity led to many inventions. In 1876, Alexander Graham Bell invented the telephone. Thomas Edison and Joseph Swan finally perfected the design of the light bulb in 1879. Contrary to popular myth, Edison did not invent the light bulb. In fact, less efficient light bulbs had been around decades earlier. So, why is Edison given credit for this invention? His design was the first one that was practical for home use. Edison's light bulb was economical and safe, and it stayed on for an impressive 13 and a half hours. Edison also provided another innovation that we take for granted today: the use of a standard light bulb base that fit into any socket.

Automobiles began to replace horse-drawn carriages in major European cities. Among the most influential and far-reaching innovations of the Second Industrial Revolution was the internal combustion engine. First introduced in 1878, it was powered by gas and air, which made it impractical for widespread public use. However, once liquid fuels, such as petroleum, were introduced, the possibilities were improved. This engine led to the development of automobiles and airplanes. Without the internal combustion engine, none of the following inventions would have been possible. In a technological leap forward, German inventor Karl Benz patented the first automobile in 1886. This was the first automobile designed to run on its own power. This model became the first automobile sold to the public in 1888. In that same year, Gottlieb Daimler invented a lightweight automobile engine. Another pioneer, Henry Ford, made his mark by manufacturing his first automobile in 1896." (study.com, 2014). Several industries and companies boomed during this era and many others went out of business due to obsolescence. Amongst the pioneers and the pioneering companies who bear their namesake that made their fortunes leading the charge during this era are individuals such as J.P. Morgan, Andrew Carnegie, Andrew W. Mellon, and John D. Rockefeller. This era also witnessed the creation of the first-ever billion dollar company on the planet, United States Steel, which was formed in 1901 by J. P. Morgan by the acquisition and consolidation of steel manufacturing companies that had been established by other "captains of industry" including Andrew Carnegie.

3. The Third Industrial Revolution (Industry 3.0)

The "Third Industrial Revolution" is said to have commenced in 1969, and is known for the age of computers and automated production. Supported by electronics and a new field known as "Information Technology", the "Third Industrial Revolution" has been undoubtedly as dramatic a transformation in world history as the "Second Industrial Revolution" was. Computers became a commonplace household item by the end of this era, fundamentally transforming how we live, study, work, and seek entertainment. As electricity and petroleum gained prominence as energy sources in the early 20th century, the "Third Industrial Revolution" witnessed the rise of nuclear energy in the second half of the century. This age gave rise to electronics, telecommunications, computers, space exploration, and dramatic advances in research and biotechnology. Computer chips, programmable logic controllers (PLCs), increasing speeds and decreasing costs, the creation and massive expansion of internet, the affordability of high-speed internet to companies and homes, robots, high-level automation that required minimal to zero human oversight, and the development of smartphones are major milestones of this era.

Companies such as Apple, Microsoft as well as products of the internet age such as Google, Facebook, Uber, Lyft, Airbnb, Tesla, SpaceX, Amazon, etc. – are all a result of this era. These companies that for the most part did not exist twenty-five to thirty years ago (or looked very different as in the case of companies that are a bit older such as Microsoft and Apple). What is also striking is that these technology companies, all products of "Industry 3.0" are dominant on the Fortune 100 list. Now consider the fact that two of the wealthiest individuals in the history of the planet, Jeff Bezos and Elon Musk, founded their companies Amazon and Tesla respectively, less than thirty years ago! To understand how impactful this era has been to each of us at this present moment, all you need to do is simply look around you. Everything that you own, everything that you interact with for work and for life – each and every convenience of life – is a direct result of inventions and innovations that emerged in this, the "Third Industrial Age".

4. The Fourth Industrial Revolution (Industry 4.0)

Welcome to the present. Welcome to the "Fourth Industrial Revolution. Depending on expert, humanity as a species entered "Industry 4.0" in 2016 or 2020. The distinction between the start year is inconsequential, but what is consequential is that as a species and society, we are just as unfamiliar with the road ahead as every other generation before us was at the start of their own respective industrial revolutions. Imagine how horse breeders and wagon manufacturers reacted with the advent of the automobile. Most likely considered this bizarre "mechanical horse" as a passing fad, or a plaything of the very wealthy. How many of these firms had the foresight to consider diversification of their organizations? How many individuals expected how their lives would change dramatically when their homes would be illuminated and warmed by electricity as compared to gaslight? If we reflect back on the three ages before this one, I am certain we can relate to how disorienting it can be to enter a new era – a fundamental shift in how we live and work, this shift driven by technology, innovations, and inventions – with absolutely no precedent to draw from.

The "Fourth Industrial Revolution" will lead to the blurring or downright dissolution of boundaries between the physical world, the digital world, and the biological world. We will inhabit a blended reality, where the real word is melded with the virtual world. This concept, the so-called "Metaverse" is only one of countless ways that our personal and professional lives are about to be as unrecognizable to us now in 2021 as it would be to our great grandparents from 1921 if they were to be dropped in the middle of an Apple Store in 2021. "The Metaverse is a combination of multiple elements of technology, including virtual reality, augmented reality and video where users "live" within a digital universe. Supporters of the metaverse envision its users working, playing and staying connected with friends through everything from concerts and conferences to virtual trips around to the world." (Snider & Molina, 2021). Described as a perfect storm of intersecting technologies – from true Artificial Intelligence (AI), AI-at-scale, robotics, Internet of Things (IoT), connected devices, 3D printing, quantum computing, this era will radically disrupt the way we live, work, study, and play. There will be no industry, company, and job on this planet that will not feel the effects of this "Fourth Industrial Revolution", the only distinction being that the level of impact of these effects might vary.

Klaus Schwab, founded and executive chairman of the World Economic Forum, and author of the book "The Fourth Industrial Revolution" (Schwab, The Fourth Industrial Revolutuion, 2017) was one of the first and most prominent thought leaders to designate this era as the start of a whole new revolution. In an article that originally was published in Foreign Affairs (Schwab, The Fourth Industrial Revolution - What It Means and How to Respond, 2015) that summarizes his thoughts on this age, Schwab writes "We stand on the brink of a technological revolution that will fundamentally alter the way we live, work, and relate to one another. In its scale, scope, and complexity, the transformation will be unlike anything humankind has experienced before. We do not yet know just how it will unfold, but one thing is clear: the response to it must be integrated and comprehensive, involving all stakeholders of the global polity, from the public and private sectors to academia and civil society. There are three reasons why today's transformations represent not merely a prolongation of the Third Industrial Revolution but rather the arrival of a Fourth and distinct one: velocity, scope, and systems impact. The speed of current breakthroughs has no historical precedent. When compared with previous industrial revolutions, the Fourth is evolving at an exponential rather than a linear pace. Moreover, it is disrupting almost every industry in every country. And the breadth and depth of these changes herald the transformation of entire systems of production, management, and governance. The possibilities of billions of people connected by mobile devices, with unprecedented processing power, storage capacity, and access to knowledge, are unlimited. And these possibilities will be multiplied by emerging technology breakthroughs in fields such as artificial intelligence, robotics, the Internet of Things, autonomous vehicles, 3-D printing, nanotechnology, biotechnology, materials science, energy storage, and quantum computing. Already, artificial intelligence is all around us, from self-driving cars and drones to virtual assistants and software that translate or invest. Impressive progress has been made in AI in recent years, driven by exponential increases in computing power and by the availability of vast amounts of data, from software used to discover new drugs to algorithms used to predict our cultural interests. Digital fabrication technologies, meanwhile, are interacting with the biological world on a daily basis. Engineers, designers, and architects are combining computational design, additive manufacturing, materials engineering, and synthetic biology to pioneer a symbiosis between microorganisms, our bodies, the products we consume, and even the buildings we inhabit." (Schwab, The Fourth Industrial Revolution - What It Means and How to Respond, 2015) Speaking to the impact of this era on businesses and people, Mr. Schwab continues "the acceleration of innovation and the velocity of disruption are hard to comprehend or anticipate and that these drivers constitute a source of constant surprise, even for the best connected and most well informed. Indeed, across all industries, there is clear evidence that the technologies that underpin the Fourth Industrial Revolution are having a major impact on businesses." For society, Mr. Schwab believes that "the Fourth Industrial Revolution, finally, will change not only what we do but also who we are. It will affect our identity and all the issues associated with it: our sense of privacy, our notions of ownership, our consumption patterns, the time we devote to work and leisure, and how we develop our careers, cultivate our skills, meet people, and nurture relationships. It is already changing our health and leading to a "quantified" self, and sooner than we think it may lead to human augmentation. The list is endless because it is bound only by our imagination." (Schwab, The Fourth Industrial Revolution - What It Means and How to Respond, 2015).

Age and Industrial Age

George Orwell is quoted as saying, "Each generation imagines itself to be more intelligent than the one that went before it, and wiser than the one that comes after it." In each of the preceding three Industrial Revolutions, there were themes - common elements – that bound that specific Industrial Age together. These common elements transcended age and generational demographic of a cohabiting workforce within one company. At any given time, reflective of society as a whole, across every enterprise, you will find young entrants who are just entering the workforce, and you will find older, senior, much more experienced personnel that are close to retirement age.

There is a broad age spectrum of workers between these two career bookends. The older workers, by virtue of their age and the fact that they have gained significant experience through their long careers, typically accrue institutional wisdom that they can impart to the rest of the organization. For the younger generations, there is as much to learn as there is to teach for the older generations. For the cohabiting cross-generational workforce, they have some precedent to draw from. Their immediate futures had been relatively predictable and controllable. Older generations were able to render career advice to all those younger than them in their careers. For the older generations, there was a shared sense of simple beliefs that their past has created the present for those who have come after them, and their collective present will set the stage for the future for those just entering the workforce. The Fourth Industrial Revolution is going to prove to be vastly different. With no precedent to draw from in terms of the rapid technology-driven changes, while the more experienced generations can provide invaluable lessons about the business and technical aspects of their industry and company, they might not be able to provide lessons about the technology and commensurate changes to their industries and the nature of their jobs. With the pace of change introducing rapid cultural and societal changes within younger generations, how they prefer to work, attitudes towards work, and how they expect to learn, each organization will be posed with an ongoing dilemma to effectively stitch together the fabric of their unique corporate culture. The one thing that will be for certain is that these distinct generations, each with their unique experiences and perspective on technology, will all be facing the common great equalizer – the unprecedented pace of change due to the exponential changes in technology.

Chapter Six – Five Generations, One Digital Planet

Much focus is put on the concept of "generations", an often vast generalization to describe people within a delineated population - born and living around the same time – who experience similar significant events, cultural shifts, and macro socioeconomic changes within a defined span of time. Despite this mass generalization, these characterizations that define traits and characteristics of each generation are by and large within the spectrum of plausible acceptability. In other words, even though these broad generalizations are often derided for developing stereotypes and stereotyping people based on the year of their birth, their acceptability in society has led to entire industries basing their product target markets on these stereotypes. Companies gear their product offerings, marketing, and advertising based on what a certain generation will respond most favorably to. Each generation, in addition to certain traits and characteristics, is defined by their outlook on culture, sense of acceptable social norms, work ethic, food preferences, fashion preferences, etc. Only in the past few decades, has each generation been starting to be defined by their inclination, or proclivities, towards technology; more specifically, how adaptable each generation is to the rapid pace of transformational technological advancements around them.

The differences between generations today, especially when it comes to the adoption of technology, could not be starker. A visible and relatable example today is how we see this generational difference play out today in how different generations have gravitated towards certain social media platforms. This association of certain age groups with certain social media platforms has certainly been unplanned. We have certainly come a long way from the early days of the groundbreaking social media platform, Myspace. Surpassing Yahoo! And Google for the mantle of most visited website in the United States in June of 2006, Myspace was the premier social media platform that dominated the "early days of the internet", from 2005 to 2008. Facebook, incepted in 2004, grew to claim a user base of 2.8 billion as of 2020 (Facebook, 2021), ranking 7th in global internet usage (Amazon, The top 500 sites on the web, 2021). The social media behemoth has seen a steady erosion of younger new users. As an article on 'The Verge' stated: "teenage users of the Facebook app in the US had declined by 13 percent since 2019 and were projected to drop 45 percent over the next two years, driving an overall decline in daily users in the company's most lucrative ad market. Young adults between the ages of 20 and 30 were expected to decline by 4 percent during the same timeframe. Making matters worse, the younger a user was, the less on average they regularly engaged with the app. The message was clear: Facebook was losing traction with younger generations fast" (Heath, 2021).

Organically falling into place, Facebook is generally considered to be the social media platform of choice for those over 30 years of age ("and of old people"), while Instagram (owned by Facebook/Meta) attracts a younger audience, whereas Snapchat and TikTok have gained immense popularity with teenagers and those in their 20s. Even within the same household with two parents and two children, it is likely that parents (and grandparents) could be users of Facebook while the children could exclusively use TikTok or Snapchat. Things can get all the more interesting when the two children have at least a three to four year age gap between them. In this scenario, it is foreseeable that while the parents leverage Facebook for social media, the older child could be exclusively an Instagram user, while the younger sibling might be an exclusively Snapchat or TikTok user. Considered in totality, how remarkable is the fact that NONE of these commonplace and highly popular platforms even existed twenty years ago? As astonishing it is to examine the gap in technological affinities and technical capabilities that have manifested themselves within a single family unit, this gap can seem like a chasm when viewed through the lens of the professional world. Once this "corporate generational technological chasm" is superimposed onto the backdrop of the continually increasing breakneck pace of technological advancements, it becomes evident that people across all industries and professions find themselves in truly uncharted waters – a situation that has no precedent to compare to or learn from.

Setting aside the centenarians, or those close to that age (people who are 100 years old, or close to 100 years old) as outliers, for the first time in history, entering a new epoch together are five distinct generations – five unique age bands – that have had very unique experiences in their personal and professional lives leading up to this new digital era, the "Fourth Industrial Age". While the workforce across industries and sectors has always been comprised of multiple generations serving within it concurrently, for the first time in history, there are five concurrent generations within the workplace (Purdue, 2020). These five generations that are currently cohabiting the public and private sector industries across the world are:

1. Traditionalists/The Silent Generation (those born between 1925 and 1945).
2. Baby Boomers (those born between 1946 and 1964).
3. Generation X/Gen X (those born between 1965 and 1979).
4. Generation Y/Gen Y – better known as Millennials (those born between 1980 and 1996).
5. Generation Z/Gen Z (those born between 1997 and 2015).

For those surprised to see "Traditionalists" on this list as a small yet integral part of the workforce, people that belong to this age group as usually employed in the public sector (elected governmental officials), in academia, or run organizations as entrepreneurs. The public sector, governmental service, is a good example of a sector where "Traditionalists" are employed. Think about the United States Senate or the United States Supreme Court. The 46th President of the United States, Joseph Robinette Biden was born in 1942 and was elected to the office at the age of 78.

The technological gap between the "Traditionalists" or "Baby Boomers" and "Generation Z" is palpable. This gap is predicated on the simple principles that one generation had to bear witness to the breakneck pace of technological progress and innovation, while the other generation bookending the spectrum is a digitally-native generation. As expected with the age gap, the generational gap of this multigenerational workforce brings with it a rich diversity of each generation's motivations, their communication style, their worldview and how their worldview was forged. For instance, "Traditionalists" were shaped by The Great Depression, World War II, the advent of radio and television, "Boomers" by the Civil Rights movement, Watergate, and the Vietnam War, "Generation X" by the fall of the Berlin Wall, the breaking up of the Soviet Union, the rise of the internet and personal computing as well as the infamous "dot com boom and bust", "Millennials" were shaped by September 11th, the widespread availability and access to high-speed internet, and smartphones, while "Generation Z" has been influenced by events such as how the world changed after September 11th, the wars that followed, the Great Recession, and most dominantly, access to technology from birth. As technologically hardwired the two latest generations are, there are still subtle distinctions between them. As an example, while "Millennials" prefer to use technology – Instant Messaging, SMS Texting, and email as a primary means of communication, "Generation Z" prefers Instant Messaging, Social Media, and SMS Texting (note the absence of email entirely). "Generation X" and "Boomers", having to adapt to the use of technology, prefer whatever means of communication is most efficient, including phone calls, or face-to-face interactions.

According to Purdue University (Purdue, 2020) and Pew Research (Pew, 2015), the current constituency of the US workforce is comprised of "Traditionalists" at 2%, "Boomers" at 25%, "Generation X" at 33%, "Millennials" at 35%, and "Generation Z" at 5%. Aside from the underpinnings of rapid and rampant technological change, this transgenerational workforce is in itself the midst of major shifts in the coming years. Each day, an approximate 10,000 "Boomers" reach retirement age (Frankel, 2017), while 65% of "Boomers" plan to work past the age of 65 (Heimlich, 2010). The distinctions on how these five generations behave or are sized in relation to the overall size of the workforce has no bearing their influence across industries. According to a report (Ang, 2021) from the Visual Capitalist, in their inaugural "Generational Power Index (GPI) for 2021", they state that "We're on the cusp of one of the most impactful generational shifts in history. As it stands, the Baby Boomers (born 1946-1964) are America's most wealthy and influential generation. But even the youngest Boomers are close to retirement, with millions leaving the workforce every year."

The report that measures a generation's influence considers a variety of factors that fall under three categories of Economic Power, Political Power, and Cultural Power. This report states that "Baby Boomers lead the pack when it comes to overall generational power, capturing 38.6%. While Boomers hold the largest share of power, it's interesting to note that they only make up 21.8% of the total U.S. population. Gen X comes in second place, capturing 30.4% of power, while Gen Z ranks last, snagging a mere 3.7%. Gen Alpha (author's note: *those born after 2013*) has yet to score on the ranking, but keep in mind that the oldest members of this generation will only be eight years old this year—they haven't even reached double digits yet." (Ang, 2021).

As evident, there are fundamental distinctions between these cohabiting generations across our workforce today. These ratios are further skewed in other regions across the world - such as within China and India - with their large working populations of "Millennials" (Neufield, 2021). These workers - belonging to different eras - work, behave, communicate, and operate very differently from one another. Their experiences and worldview drive their ideologies, working styles, and how they interact with others within their generation and outside of it. On the surface, it would appear as though representatives of each generation have very little in common in the shared workplace of today, and the workplace of ten to twenty years from now. Beneath the surface however lies a simple shared truth that all these cohabiting generations have one thing in common. They are share the fact that regardless of whether it is a generation that has had to adapt to technology permeating every facet of their workplace like with "Generation X" or "Boomers", or those born "digitally native" such as the younger "Millennials" or "Generation Z" – none of us in any of these generations can fathom and fully appreciate how impactful and transformative the next couple of decades are going to be to our industries, sectors, companies, workplaces, and our personal lives.

None of us across any generation can imagine the disruption in store ahead for us and the resulting career opportunities - and career perils - that this disruption could bring. The source and root cause of this disruption and transformation in the next twenty years will be the same as it has been in the past twenty. If the past twenty years are any indication what is going to unfold in the decades ahead, it will mean massive change for industries, companies, and careers as a direct result of technology and digitization. The sole distinction between the past twenty years and next twenty years is that the technological change twenty years from now will make the technological progress of the past twenty years seem analogous to comparing a typewriter to a modern PC and word processing application – both are founded on the same concepts, but there really is no equivalence between the two.

Technology has undoubtedly positively influenced the quality of life for hundreds of millions. It has also brought along with it great cultural and macro socio-economic challenges (as well as arguably geopolitical changes due to social media being weaponized to spread disinformation). Those who have had their jobs disrupted or eliminated in the past several years due to digitization and automation look upon technology in the workplace very differently than those jobs technology has created. Those "Boomers" who are on the cusp of retirement and are threatened by job disruption in the next few years due to technology have a very different perspective than an individual belonging to "Generation Z" who is just entering the workforce in a job that did not exist or looked radically different when the "Boomer" entered the workforce. However, what the "Generation Z" worker of today needs to clearly understand is that their career has just as many opportunities and faces just as much peril in the next few decades as the "Boomer" of today does, as does an individual who belongs to "Generation X" or a "Millennial". Should we as workers, regardless of industry, vocation, or generation not be astutely aware of these rapid digital changes around us, we will be unable to capitalize on the opportunities that will abound. As unfortunate as it is to miss abundant opportunities, a more pressing concern is that we, regardless of generation, will be unaware of the risks posed to our careers and continued gainful employment by the technological changes and digitization.

To Understand Is To Prepare, To Prepare Is To Understand

The unprecedented and unforeseeable very nature of the "Fourth Industrial Revolution" can generate feelings of anxiety and uncertainty about our futures. These concerns are centered around our careers and the implications of rampant and rapid digitization on our careers. We like to be in control of our careers. We want to be assured that regardless of if we are just starting out in our career, somewhere in the middle of it, or a few years shy of our retirement, we are going to be able to continue our contributions - and earn appropriate, uninterrupted compensation from it. Whether you are a doctor, an electrician, a lawyer, a software engineer, a teacher, or own a business, you want to add value, feel valued, and derive value from work.

Entering the "Fourth Industrial Revolution", we – all of us, all workers across all professions regardless of generation we belong to – are at a unique time in history. We collectively stand on the entryway of a new industrial age while walking out of the exit door of the previous one. It's akin to walking from one room into another. The room that you are leaving is familiar to you. You've been present in that room for a little while. It is fully illuminated, and the corners of the room that you don't see are irrelevant to you. The room that you are walking into is one that is completely unfamiliar to you. You have never been in this room before, and the interesting part is that neither has anyone else. No one can tell you the dimensions of this room. You might as well be blindfolded as you leave your familiar, illuminated room and walk into this new one, taking a leap of faith that you will be able to get around in this unfamiliar, new room.

Now imagine walking into this new, unfamiliar room having been provided with the schematics of this room – the dimensions, layout, etc. The room will still be unfamiliar, but you will have a blueprint to stride into this unfamiliar room and navigate yourself around – aware and in control. That is what we are going to do over the next several chapters – help equip ourselves with the knowledge and tools to survive in some scenarios, thrive in others, and truly "**thurvive**" in this revolution. Knowledge, planning, and preparation will help us gain confidence and convert any feelings of anxiety and uncertainty into excitement and anticipation. So where do we go from here and how do we gain the knowledge we need to prepare for this Digital Age? We have reviewed the cohabiting generations and gained an understanding and appreciation of the three "Industrial Revolutions" that have come before the one we are entering, along with the disruptions that these can trigger.

We will explore some of the disruptions we might expect in the next decade or two, how these might create opportunities for us, what risks or threats our careers and companies might face, and how we can successfully navigate these unchartered waters. Before looking inwards and helping ourselves get ready for the "Fourth Industrial Revolution", as unprecedented as this era is anticipated to be, there are generally familiar ways in which companies and industries manage change. These reactions are observed corporate/industry behavior and are categorized into six groupings. These groupings will likely provide a sneak peek, a preview, into the shape of things to come from companies within our industries. These actions have clear and direct implications to our jobs and careers, and in advance of looking inwards and understanding how we can prepare ourselves for "thurviving" the next several decades of this new era, it is helpful to understand how organizations might respond and react to these changes. From this perspective, we will also delve into corporate and organizational behavior in times of rapid change.

"**Thurvival**" in this Digital Age will require us to transform the way we think and reprogram ourselves. We will need to adopt and adapt to changes faster than we have ever before, while also keeping an open mind that we might need to adapt to macro changes multiple times over our careers. Drawing from a quote that the American futurist Alvin Toffler is credited with from his book "Future Shock" (1970), "The illiterate of the 21st century will not be those who cannot read and write, but those who cannot learn, unlearn, and relearn". We will understand the concept of "Learn-Unlearn-Relearn" as it applies to our careers, and how we can put in the work to future-proof our careers regardless of what happens around us. This is where the work that we have to put into it comes in. We will need proactively prepare for a future that is not entirely clear to us, and rewiring our own innate way of thinking while doing so. As a transitionary generation, we are - all of us - starting at somewhat of a disadvantage.

We are disadvantaged by not having the benefit of lessons from a few decades of going through a new "Industrial Age" behind us. We are disadvantaged by not having access to historical precedent. We are disadvantaged by not having adequate data about the past, and being unable to forecast the future. But most of all, we are disadvantaged by our own internal "hard wiring" – the greatest of all our aforementioned challenges. Our internal "hard wiring" that makes us averse to change, especially when we have little to no control over this broader, macroscopic change. It is our resistance and aversion to change – anything that we perceive is "being done to us" - that is crucial for us to get past in order to adopt and adapt. Our goals are to overcome these disadvantages, and we are absolutely more than capable of doing so.

We generally think that change that is done **to** us is always bad, and that change that is done **with** us is always good. It is this perception that prevents us from being able to adopt and adapt as swiftly as we need to at times. This ability to rapidly adopt and adapt is more critical now than at any other time in the past. It is important to understand why these deep-rooted perceptions regarding change exist in order to successfully be able to shift our mindset. We will explore the concept of change and our instinct to resist any forms of it over the next few chapters. Whether you are twenty five years old or fifty two years old, with this understanding serving as a foundation, we can then begin our vital journey on learning how to "**thurvive**" in the "Fourth Industrial Revolution".

SECTION THREE

UNDERSTAND THE NEED TO WANT TO CHANGE

The Past and the Present to Prepare for the Future

Chapter Seven – The Binary Choice

"It is not the strongest nor the most intelligent of species that survives, but the one that is most adaptable to change." (Darwin, 1859) – Charles Darwin.

"It is not the most intellectual of the species that survives; it is not the strongest that survives; but the species that survives is the one that is able best to adapt and adjust to the changing environment in which it finds itself" (Megginson, 1963) – Dr. Leon C. Megginson.

Consider these two famous and often quoted statements, one seemingly building upon the other. At face value, the distinction between these two statements seem to be those of mere semantics. Dr. Megginson seems to be simply expanding upon Darwin's original quote. But within these semantics resides a deeper message that provides clarity and specificity. Darwin's original statement states that those species that change and adapt more rapidly are ones that are more successful. Dr. Megginson's statement extends Darwin's quote, stating specifically that those species that change and adapt more rapidly as a response to *external stimuli from their environment* are the ones that find greater success. The external stimulus that we are all experiencing is technology, more specifically, the rapid pace at which technology is itself evolving. Those who are able to change and adapt to this external stimulus of technology, are the ones who will realize greater success – professionally and personally. If there is one word to describe this accelerated and accelerating technological era that we are all in, it would be - "**change**". No matter what generational category you belong to, and no matter how vastly dissimilar each category is in their much-stereotyped generational traits and behaviors, the one commonality that binds all generations, is **change**.

"Change". What is it about this word that inspires strong feelings within a broad majority of people? Whether these feelings are positive or negative, the word "change" is seldom met with a reaction of abject apathy. People's reactions to this word spans the spectrum of reactions – some claim to love change and seek to embody a continual and incremental improvement mindset, while others detest change, and shun the very idea of having *to* change. For the most part, most of us state that we keep an open mind to change, but are inconsistent in our adoption of change. People, especially in a corporate setting, can say they *want* change, but can be resistant *to want to* change.

"If you don't like change, you'll like irrelevance even less." – General Eric Shinseki, former U.S. Army Chief of Staff. I am unsure if General Shinseki realized it when he said these words, but there are very few quotes more fitting for the "Fourth Industrial Age" than this one. Change is often considered a binary choice – change or stagnate, change or let change happen to you, change or become irrelevant, or "Change or Die" as author Alan Deutschman's book (Deutschman, 2007) title stated. And herein, within this binary choice, lies the paradox of the 21st century. Thinking in absolute binary choices restricts us to having to choose between one of two choices: "change or don't change", "on or off", "black or white", "success or failure", or "1s or 0s" (in a digital sense). Thinking in absolute binary choices presents to us a false illusion that we actually have a binary choice. We don't have a choice but to start preparing for a future which will present significant opportunities and some significant risks due to technology. We need to ready ourselves to capitalize on the opportunities and prepare ourselves to stave off the risks. Our success in the "Fourth Industrial Revolution" will operate without the constraints of these absolutes – because we need to clearly understand that a binary ***choice*** simply does not exist. In an era which will require us to be open to constant change and demonstrate the ability to rapidly adopt change and adapt to changing circumstances, the greatest challenge we need to **survive** is our own hardwired resistance to change. Our goal will be to maximize our opportunities to **thrive**. That is where the concept of "**thurvival**" originates – pushing through the uncertainty over the next several decades and overcoming the lack of historical precedent. "**Thurvival**" requires us to understand that the change around us is not something we can control, but what we can control is how we individually respond to it in order to capitalize on the opportunities ahead.

Change is currently running rampant and it's currently running rampant all around us. One might successfully argue that this change that has been occurring all around us has been exponentially accelerating over the past two to three decades. The false illusion of choice that we had been historically presented with for most things in life has likely only postponed the inevitable. Consider the fact that twenty years ago, most of us retained a choice whether to give up our home landlines in favor of exclusively using mobile phones. Most who resisted that choice, at some point over the past five years have inevitably abandoned landlines for their mobile phones. The same holds true with those who resisted moving from a flip phone to a smartphone ten years ago – they may have resisted for the longest time, but at some point had to embrace the inevitable.

There is likely a miniscule minority that continues to resist this transition, but they cannot likely holdout forever. Consider that even today most of us retain the choice whether to "cut the cord" and discontinue service from our local cable TV providers in favor of purely streaming services. The simple fact remains that when it comes to technology and our societal technological revolution, we do not have a choice. We might be able to delay and postpone the inevitable, we might have a choice in not upgrading to a new phone for a few years, or not subscribing to every streaming service there is, but our choices on whether to digitize or not, simply do not exist. It's more a matter of time than it is a matter of having a binary choice. And once again, consumer electronics and services are a good proxy for what will transpire in our professional lives as well. There is simply no choice. "**Thurvival**" is a necessary skill.

Before we learn about how to "**thurvive**" in this digital age, it is important to understand the roots of our resistance to change, and the importance of consciously shifting our own mindset on our reaction and response to change.

Change - Currently Rampant All Around Us

Where there are manual, rote, operational tasks, there is potential for significant disruption due to automation and AI. If you close your eyes and imagine the rote, operational, "hamster wheel" type tasks that you perform within your own job, you can envision what aspects of your own profession are likely to be subject to automation and technological disruption. You can likely develop a pretty comprehensive list of such manual tasks across a myriad of professions if you think about people you might encounter on a daily basis and think about their jobs. From there, you do not need to be especially intuitive to extrapolate a list of jobs that will continue being eroded due to AI and automation just by the end of this decade. What is a bit trickier to intuit is that when change comes, it comes quickly. What's even trickier is to recognize the rapid transformation of entire industries that might be occurring in real time right under our noses. One of the challenges in being able to recognize current and ongoing disruption within an industry is because we are living through it. Another challenge is that there are sectors of industries that we cannot witness direct change taking place. These sectors interact with as consumers only indirectly, and therefore any changes within these industries are buffeted by layers of the supply and ecosystem value chain.

Take the simple example of photography film and being able to develop photo prints from film rolls today. Film development used to be a common service offered at most major retailers such as Walmart and almost every drug and pharmacy store such as Rite Aid, Walgreens, CVS, etc. As we looked at the example of Kodak earlier in this book, with the rise of smartphones and the precipitous decline of cameras and physical photos being developed, film development is a service that is not offered at any of these establishments today. While one can still purchase disposable cameras at these retailers, the film contained inside has to be dispatched to a special studio to be developed. This change happened seemingly overnight, but goes to prove that when one domino in an ecosystem falls – in this case, photography being disrupted – every other associated industry and company within that ecosystem swiftly feels and experiences these reverberations.

Consider the example of restaurant servers and wait staff. Whereas most sit-down restaurants today continue to focus on wait staff as a key part of the experience they are seeking to provide, wait staff that served as intermediaries in establishments that aren't quite drive-through type fast food establishments and not quite sit-down restaurants, have shifted away from being a waiter-dependent business. Airport restaurants are a prime example of this marked shift. Sit-down airport restaurants used to offer services commensurate to any other sit-down establishment, albeit with a focus on quicker service and a generally "frills-free" dining experience. A waiter used to take your order and ensure your dining needs were met as you jetted off from one destination to another. Through the pandemic, where the reliance on wait staff posed unique challenges, these restaurants accelerated a move that was already underway. Today, it is more common than not to sit at an airport restaurant, whip out your smartphone and scan the QR code for a menu, place and pay for your order online, and have a server simply bring the meal to your table. In some restaurants, they have done away with the server entirely, and you are expected to walk up to the counter to pick up your order. Some other restaurants have replaced a waiter with a robot server who brings your food and beverage order out to your table. The underlying technology has allowed these restaurants to streamline their operations, reduce operating costs, provide quicker turnarounds, and leverage technology to provide an experience that most travelers seek – a quick, hot meal with as little interaction as is required. From being able to check government issued ID for validating someone's age to serve alcohol to being able to settle your tab – everything can be performed in a relative touchless manner with little to no need for human interaction. Most restaurants increased and accelerated their adoption of technology during the pandemic when sit-down dining was not possible due to quarantine restrictions, and delivery services ensured that they continue staying afloat.

Unless you've been a regular business traveler, it is easy to overlook technological advancements such as the ones with airport dining that are all around us. Going deeper into the supply chain, my opinion is that the most often overlooked technological advancements that diners at the airport and the broad population at large – each and every individual, from the youngest to the oldest – take for granted all day, every day, is in the agricultural sector. Farming is the world's oldest industry. Farming allowed humans to settle and build communities, communities that developed into settlements, villages, towns, cities, and countries. The shift of humanity from a nomadic hunter-gatherer settlement to established communities where the local land provided sustenance began around 12,000 years ago when the first lands began to be farmed. This period, termed as the "Neolithic Revolution" allowed for reliable food supplies. Predicated on the fact that crops and animals could be farmed to meet food demand, the world's population has grown from around five million people 10,000 years ago to a whopping 8 billion people today.

For the longest time throughout our history, sustainable farming within a community meant the difference between starvation and sustenance. A bad year for crops, the lack of rainfall, pests, diseases, or anything that would adversely impact an expected yield were harbingers of challenging times for many families through those farming communities. As cities expanded and settled, and modes of distribution, storage, transport, preservation, and refrigeration became commonplace, farming has become relegated to farming communities and agricultural states. Unless designated as local produce, or "farm-to-table", the vegetables, fruit, or meat that an average American family consumes, could have been cultivated anywhere on the planet. Most people do not pay heed to where their produce originates. We might glance at the labels that designate country of origin for our avocados, or favor meat that originates in the Midwest, but by and large, the average consumer does not know, and more specifically, does not care. Most people seek quality products at an affordable price. That is the extent of an average consumer's interaction with farming today. The average consumer 200 years ago likely knew the farmer by name, or as was typically the case, the food they consumed were cultivated by their own family, and it likely had involved brutally hard work on their own part to get that food to their plate.

Farming was difficult a hundred plus years ago. Farming still – to this day – requires determination, grit, commitment, and a level of persistence that most people – by virtue of being able to walk down the aisle of their grocery store, or order their groceries online – are oblivious to. Agriculture and farming – where people are actually physically tending fields or animals – is still a demanding profession. The farming ecosystem today is occupied and at times dominated by the big agricultural organizations such as Cargill, Archer-Daniels-Midland, Bayer, etc. These firms and the agricultural ecosystem has invested a significant amount in digitization and modernization in order to assist farmers, improve yields and have predictable outcomes of high quality. Whether it is using AI to bioengineer specific crops to be resistant to specific pests, or develop hybrid vegetables that might be hardier, or use less water, or be more tolerant to climactic extremes, agricultural technology, or agtech, is big business today. It is also one of the largest and most fundamental sectors of our economy that we take for granted. And since we don't think of agriculture and farming every time we step into a grocery store, or order food at a restaurant, it is easy for us to be oblivious to the fact that technology disruption has not left this ancient industry untouched either.

While it is obvious that technology has only served to better farming and agriculture, it is indisputable that tasks that used to leverage humans in the past are automated by technology and AI today, a trend that will only accelerate and continue. A 2015 report by McKinsey & Company characterized farming as the world's least digitized industry (McKinsey Global Institute, Digital America: A tale of the haves and have-mores, 2015). If you consider all the agriculture in aggregate across the world, this characterization makes sense. There are developing countries, or major swaths of developing countries, where partial industrialization exists, but farming is still a primary industry. With less financial resources available to these agricultural economies, there are limited, equitable opportunities for individual farming communities to digitize. Less than two years after their study, McKinsey & Company released another report (McKinsey & Company, 2017) that conducted an assessment of the potential for automation across several industries. This 2017 study listed agriculture as fourth with a 57% potential for automation, behind accommodation and food services (73%), manufacturing (60%), and transportation and warehousing (60%), but ahead of retail trade (53%), mining (51%), construction (47%), and finance and insurance (43%), etc.

Farming in the more industrialized economies is already undergoing a significant technological revolution, and although these changes aren't as visible and prominent as Netflix usurping Blockbuster, or Uber disrupting the world's taxi cab industry, they are just as vital to reinforce the fact that digitization will touch every single sector and industry. AI to facilitate decision-making, sample soil and climatic conditions and provide guidance to farmers on optimal yields, pest management; robots to harvest crop and cultivate/fertilize soil; drones to aerially monitor hundreds of acres and provide data back to farmers; IoT and sensor devices embedded across devices and even animals; the use of Blockchain to track origin and provenance of food (critical for traceability of livestock, such as with poultry during the Avian Bird Flu); data analytics; etc.; are only a few of the technology-fueled disruptions underway today in farming and agriculture. Remember, less visible to us does not mean that it is not happening – no industry is insulated.

Change, The Only Constant

In his book entitled "Fragments", the Greek philosopher, Heraclitus of Ephesus, said "change is the only constant" (often quoted as "the only constant in life is change"). Heading into the third decade of the 21st century, it is highly likely that every worker, in every company, across every industry, around the planet, uses or interacts with technology in some form or fashion in order to do their job. Some industries are more technologically progressive than others, the same being true for sectors within a particular industry, and even companies within a particular sector. Even those firms and sectors that have traditionally been thought of as laggards in digitization - and slow in making earnest investments in technology - have made sizeable efforts in an effort to catch up over the past decade. However, like with anything new, the introduction of digitization at scale and the implementation of technology to augment people's jobs has not always been well received.

Potentially being perceived as disruptive, employees have been passive, or have outright resisted technological adoption in their workplaces. For some, it has been about "this is the way we've always done it", and for others, it has been received as something done onto them without their consent, buy-in, or having solicited their input. The pace of progression and implementation of technology in and of itself is meaningless if there is no commensurate adoption of it within the workplace. The best laid digital strategies are not focused as much on technology, as they are on people, and how organizations can successfully manage the change curve. The concept of workers resisting new technology is not a 21st century problem, or even a 20th century one.

The term "Luddites" is used to describe British workers in the middle of the First Industrial Revolution in the early 1800s. Luddites were textile mill workers and weavers who were said to have been so opposed to new technology, and the perceived disruption of their source of income by introduction of machinery within cotton mills, that they destroyed these machines. Luddite is now used as a blanket term to describe people who dislike new technology. Depending on situation and person, this term can be used as a badge of honor and a term of endearment, or as a derogatory term impugning someone's technological ineptitude. Even if we would not consider workers throughout the latter half of the 20th century and into the 21st century as "full-blown Luddites", there are a significant number of workers, regardless of industry or profession, that demonstrate technology resistance at varying degrees. This spectrum ranges from all out hostility to technological progress, to apathy towards using new tools, to being amenable to how they can augment their jobs, to the "soft torpedoing of technology" via underutilization. Workers across laggard sectors, by-and-large, have made adoption of technology much slower than it could have been. In doing so, these workers inadvertently risk their own company's futures, and implicitly, their own jobs. If a vast majority of workers across many firms exhibit similar "dig in their heels" behavior, it can truly set an organization back behind the technological curve.

While "technophobes" and the "technically challenged" have been around throughout the past three Industrial Revolutions, the latter half of the "Third Industrial Revolution" and the start of the "Fourth Industrial Revolution" poses some very unique challenges, and reveals a dichotomy that is unique to our times. I would not go as far as to label workers who are intentionally reticent against adopting technology within their workplace as "21st Century Luddites" or "Neo Luddites" as they are commonly referred to today, but it would be fair to call them "Technology Resistors". The unique challenges of this era that "Technology Resistors" face in contrast to the "Luddites" of the past is that we, as workers today, simply do not have time to fall behind. Falling behind technologically, given the breakneck pace of change around us, is a recipe for certain doom for a firm and poses a grave threat to the careers and gainful employment for people who fail to adapt. The interesting paradox even with "Technology Resistors" is that they are not averse to all technology. The dichotomy herein is that the same people who resist technology at work are also the ones who have the latest and greatest consumer electronics in their homes. How bewildering is it that the same people who push back against new tools and technology in the workplace are the ones who have the latest gizmos ranging from the newest iPhone to the newest Amazon robot? Even the ones who are not "early adopters" of personal-use technology, are themselves quite well equipped with relatively recent technology releases. They might have an iPhone or a Samsung device that is four years old, but are quite astute and proficient at using these devices. Therefore, they could not be considered as "technophobes" or "Luddites" in the truest sense.

So then, why do individuals, who are otherwise quite technically capable (if not downright competent) in their personal lives develop recalcitrance to technology in their workplace? For some who are not technophobes, it could be that they do not understand technology in context of their job, and do not understand how impactful it is to them and their futures. Others yearn for a simpler time and their desire to free themselves – even for a little while – from an always-connected digital future. This can manifest itself as unintentionally being lethargic in the adoption of technology at work. And who would blame anyone for yearning for a slower-paced, less digital, less always-connected life? We have been leveraging a period of twenty-five years as a yardstick of the pace of digitization extensively throughout this book, whether to look forward and imagine a digital world twenty-five years into the future, based on a digitizing world twenty-five years in the past. Imagine being in the shoes of some of these individuals who did not ask for or choose for technology having become in increasingly inexorable part of their lives.

Imagine being in the shoes of some of these individuals who did not ask for or choose for technology having become in increasingly inexorable part of their lives. No matter how real the very benefits and conveniences they realize on a daily basis, things might have progressed far too rapidly for them to wrap their heads around it. Consider that twenty-five years ago, the internet was an escape from reality and the real world. Today, reality and the real world is an escape from the internet. Tomorrow, reality and the internet will be intermingled together into the "Metaverse", and there truly will be zero escape from the digitally connected world around us. People do not feel relaxed on vacations today unless they head to a destination with poor cellular coverage. Think about the fact that the term "unplug" did not exist in the way we use it today, and when it was used twenty-five years ago, it was most often in reference to whether you remembered to unplug the hair dryer when you left for work (or carried more somber connotations when used in a reference to human beings hooked up to life support systems).

When we think about the constant state of change we find ourselves in, it is easy for us to focus internally - on ourselves. How a change or a series of changes impact us, our livelihoods, and the way we live. This is a natural response to change via external stimulus. The human brain is hard-wired to keep us alive, not to keep us comfortable. When faced with change, we look inward, trying to understand its effect in order to take some type of an action. Often this action is just slow reaction or abject inaction – both of which are quite detrimental to the cause of "**thurvival**". What we do not do enough of is to zoom out and get a broader perspective of the macro effects of the change around us, on our culture, and on our society. Sometimes changes are fleeting and are relegated to history as passing fads, and other times, these changes are lasting and set such deep roots so quickly that they become inexorably intertwined into every single facet of our society. Technology isn't simply only influencing a cultural or societal change – it is transforming us, fundamentally and forever, as a species.

Chapter Eight – A Species, Forever Transformed

When we step back and take a macro view of technological evolution, it is indisputable that despite how people feel about one specific technology or another, technology overall has dramatically and significantly positively improved human life. Our standard of living and quality of life have indescribably benefitted by the use of technology to solve everyday problems, automate tasks, making it easier to perform daily tasks, and deliver the modern conveniences that our ancestors would envision as the stuff of science fiction. Virtually every aspect of human life today is influenced and likely significantly bettered due to technology. Technology has brought the world metaphorically closer together, and although it might seem like paradoxically some technology has moved humans further apart.

While it is easy for us to focus on technology's impact on our own lives and how technology enables our own needs, every technological development – no matter how seemingly miniscule – plays a part in bringing about societal change in some shape, form, or fashion. Some technology successfully triggers visible societal changes such as smartphones or social media in the early 21st century, or electricity, automobiles, telephones, airplane travel, computers, and the internet in the 20th century, other technology had limited impact such as travel by the Zeppelin airship, the Betamax, or even the Google Glass. However, all-in-all, there is no technology, whether a resonant success or an abject failure, that has had no impact on human society. The most interesting part of new technological development and societal change is that one feeds the other in somewhat of an infinite loop.

All technological innovations are developed in an attempt to meet an unmet need in human society, whether that need is anticipated or unexpected to the broader population. For instance, no one anticipated that smartphones would come to become a necessity across the population in such an extraordinarily short span of time, while technological progress in refrigeration and the invention of the refrigerator clearly met a desired need within human society. How remarkable is it that in the early 20th century, ice – commonplace in most household kitchen's icemakers today – used to be harvested in the winters, stored in "ice boxes", and transported to customers in the summers. Once a technology is introduced into society, it has the propensity to impact, influence, and alter a society's behavior. This behavioral shift changes culture and how a society sees and interacts with the technology. This shift in turn creates new opportunities for technology innovation, surfaces new problems that additional innovation can solve for, creates new ways in which societies operate, and can fundamentally alter human lifestyles. This cycle keeps repeating in a never-ending loop, and interestingly enough, these innovations keep seeking to "level up" by solving increasingly complex problems.

The need to solve for increasingly complex problems requires much more complex technology, which in turn then needs much greater computational needs. These computational needs then need to support computer calculations that far exceed human capacity and therefore lead to the creation of machines that think for themselves – therefore, AI. This infinite loop of innovation will itself then be elevated over the next several decades, predicated on "Artificial General Intelligence" (AGI), to assume levels of complexity that sound as farfetched to us as of the early 2020s decade that the concept of ubiquitous electricity and indoor plumbing might have sounded to human society in the early 1820s. What we just don't know quite yet is how humans will evolve commensurate to this technological innovation. If Kurzweil's postulations prove correct, humans and technology will be inexorably connected by 2045.

This societal and cultural shift has been highly visible over the past few decades. Think of how we work today versus how just a generation before us worked. Think of how we live today versus how just a generation before us lived. Think of the intermingling of work and life today and how those were distinct just a generation prior. The concept of a "work-life balance" came to use in the 1970s and 1980s (Alton, 2018) as workers sought to strike a balance between their careers, families, and other facets of life. It is easy for us to become enraptured into the packed schedules and the repeatability of everyday life. We generally adapt to our schedules, as busy working parents today try to balance their careers, shuttling their kids off to school and after-school activities, social engagements, and so on. We might become stressed about this lifestyle, bemoan the lack of downtime, or even make concerted attempts at truly relaxing and recharging.

As a matter of fact, if you are over the age of 35, you will vividly recall a time when life was not as busy, or as tiresome. Some of this feeling of always being busy is a natural side-effect of growing up and taking on more responsibilities ("adulting" as it is known in pop culture), but just over the past 25 years, the amount of "mindshare" a person can allocate to specific things has continued to diminish. We are so heads-down being busy that we haven't really taken a step back to think about why – why we are so taxed for time, why is it that our schedules are planned hour-by-hour and in some cases, minute-by-precious-minute. Being busy from the time you wake up to the time you hit the bed is not a new phenomenon. People a hundred years ago were generally certainly busy sunrise to sunset and could successfully argue that life was an order of magnitude tougher back then. One could also successfully argue that in the 21st century, we actually have it an order of magnitude easier than people a generation or two before us. Consider people who toiled on a farm two generations ago. Individuals without access to automation, electricity, irrigation, and industrial farming equipment. There are, of course, millions of people around the world who continue to have access to only some, not all, of these modern advancements today. These are incredibly tough and physically taxing jobs and require the workers serving in these vocations to work from dawn to dusk for, at times, to be able to afford bare minimum sustenance. And yet what is it about the nature of our work and lives just in the past 25 years feels very different to all? Why is it that we cannot seem to take a minute to catch our breath?

The answer is that because only in the past 25 years have we, as a society, been always "plugged in". There is no separation of work and life for millions of people serving in professional jobs. Even those who serve in skilled trades are always inundated with data and information, either by virtue of their connected device tools, such as with their foremen sending additional instructions, or even a dispatch for a plumbing company reaching out off-hours via SMS text and email for an emergency job. We are, each of us, constantly barraged by a deluge of data and information on a 24 x 7 x 365 basis. Those born in the 1980s and earlier will recall a time when personal computers were not a regular fixture in most homes. The arrival of Windows 95 and the consequentially easier access to the internet via a host of internet browsers – Internet Explorer and Netscape predominantly – changed that paradigm. "Getting on the internet" in its early days was not an easy task either. Most will have vivid memories of the screeching modems and dial-up internet, with speeds that we once considered acceptable, but in today's world, would be highly unacceptable. Consider the term "getting on the internet" for a moment. In just twenty-five years, we have gone from "getting on the internet" (and potentially getting disconnected if someone had to use the home phone – a landline – to being always connected to the internet with always available high speed broadband wherever you are (for the most part) on your mobile phone. How constantly buried are we in our phones? As Image 12 below demonstrates, we are constantly buried in our phones enough to trigger some countries such as Germany, Sweden, Korea, etc. to alert motorists of pedestrians who might be distracted by their phones and therefore might not be paying attention to crosswalks, thereby endangering their own lives while cross a street.

Image 11 – Street signs warning motorists and pedestrians about smartphone users who don't look up while crossing the road

Can we envision our lives without the internet today? Extending that thought, can we imagine our lives without our smartphones for any length of time? For a majority of people, smartphones are the first thing we look at when we wake up, and the last thing we look at when we go to bed. According to a 2020 survey by SimpleTexting, a leading platform for mass text message campaigns, "surveyed hundreds of U.S. consumers, business owners, and marketers to find their preferences and motivations for texting and text marketing", "42% of Millennials check their text messages 10 or more times per day on average. When asked, 'how quick are you to check your text notifications throughout the day,' almost 60% of all consumers surveyed answered within 1-5 minutes of the text coming through." (Wilkinson, 2020). Another survey by the same firm that was sent to 1000 smartphone users, across demographics, across 50 states pondered what people would sacrifice to keep their smartphones. The results were interesting to say the least. According to the survey results, 40% of respondents would rather be separated from their dog over their smartphone for a month, 44% said the same about their significant other, 72% would rather be separated from alcohol, and 64% stated that they would rather be without coffee for a month than their smartphone (Tocci, 2020). The 2018 version of the same survey found that nearly two-thirds of

"Millennials" would sacrifice their sense of smell and 10% would rather lose a pinky finger than give up their smartphones.

How inexorably are our lives intertwined around the internet today? Consider the following statistics on what happens on the internet in each single minute in 2021. Courtesy, Lori Lewis and All Access, here is each minute on the internet in 2021 (Lewis, 2021):

- 21.1 million text messages sent every single minute
- 1.4 million people are scrolling Facebook every single minute
- 500 hours of content is uploaded to YouTube every single minute
- 414,764 apps are downloaded from the Apple App Store and from Google Play every single minute
- 695,000 Instagram stories are shared every single minute
- 200,000 people are concurrently tweeting every single minute
- 197.6 million emails are sent every single minute
- 5000 downloads from TikTok every single minute
- 3 million images are viewed on imgur every single minute
- 69 million messages are sent via WhatsApp and Facebook Messenger every single minute
- 3.4 million Snapchat snaps are created every single minute
- $1.6 million is spent in online shopping every single minute
- 9132 new LinkedIn connections are made every single minute
- 28,000 people are watching Netflix every single minute
- 2 million swipes happen on Tinder every single minute

Domo, a cloud software company that specializes in business intelligence tools and data visualization, produces an annual "Data Never Sleeps" infographic. The infographic seeks to depict how much data is generated in a "digital minute". According to the 9th edition of this annual infographic, "Data Never Sleeps 9.0" (James, 2021):

- 167 million videos are watched by TikTok users every single minute
- 5.7 million Google searches are conducted every single minute
- 12 million people send an iPhone iMessage every single minute
- 2 million Snapchat snaps are sent (versus 3.4 million created as above) every single minute
- $283,000 is what customers spend on Amazon.com every single minute
- 6 million people concurrently shop online every single minute
- $304,000 is sent by Venmo users every single minute
- 148,000 messages are sent by Slack users every single minute
- 856 minutes of webinars are hosted by Zoom users every single minute
- 100,000 users are connected via Microsoft Teams every single minute
- 452,000 hours of media is streamed by Netflix users every single minute
- 694,000 hours of media is streamed by YouTube users every single minute
- 44 million views is what Facebook Live receives every single minute
- 240,000 photos are shared by Facebook users every single minute
- 65,000 photos are shared by Instagram users every single minute
- 575,000 tweets are posted by Twitter users every single minute

The same infographic states that "as of July 2021, the internet reaches 65% of the world's population and now represents 5.17 billion people – a 10% increase from January 2021. Of this total, 92.6% accessed the internet via mobile devices" (James, 2021). The graphic goes onto state that "the total amount of data consumed globally in 2021 was 79 zettabytes, an annual number projected to grow to over 180 zettabytes by 2025." (James, 2021). For contrast, ONE zettabyte is equal to a thousand exabytes, a billion terabytes, or a trillion gigabytes. One zettabyte is roughly equal to 1125899906842600 megabytes (MB). 79 zettabytes equates to 88946092640567000 megabytes (MB) - or 8.8946092640567 × 1016 megabyte in scientific notation. Now think about the fact that each floppy disk in the late 1990s to the early 2000s was 1.44 MB. Remember that image of the IBM 5 MB storage disk being transported in the late 1950s?

We used this analogy for the number of photos humanity takes every single moment, but it's a good reminder to understand how unimaginably large a zettabyte is. Megabytes, gigabytes, and terabytes seem quite arbitrary to most people. They are as arbitrary to materially quantify as the numbers of millions, billions, and trillions. However, since everyone knows and understands the construct of how to measure time - seconds/minutes/hours/days/weeks/months/years, and so on - as an analogy, let's consider quantifying these numbers in the time scale. A single day on Earth, 24 hours, converts to 86,400 seconds. A million seconds is 12 days. A billion seconds is approximately 32 years. A trillion seconds is approximately 32,000 years. Attribute one byte to one second, and in that metaphorical sense, when you attribute that a trillion bytes is a terabyte, you can begin to fathom the incredibly large amounts of storage we are talking about.

A report in the Visual Capitalist (Ali, 2021) provides a good barometer of the growth of these popular services over the past five years: "With the amount of data and information in the digital universe effectively doubling every two years, it's fair to say the internet minute has gone through some changes over the years. Here are just some areas that have experienced impressive growth:

- In 2016, Snapchat users 527k photos per minute, compared to 2 million in 2021
- In 2017, Twitter saw 452k Tweets per minute, compared to 575k in 2021
- In 2018, $862,823 was spent online shopping, while 2 million people were shopping per minute in 2021

- In 2019, 4.5 million videos on YouTube were being viewed every minute, while in 2021 users were streaming 694k hours
- In 2020, Netflix users streamed 404k hours per minute, growing to 452k hours in 2021"

Dissolving The Work-Life Lines

While the statistics of internet-based technology usage above underscore how indelibly the internet exists in our everyday personal lives, the always-on, always-connected nature of the internet has fundamentally transformed our work lives. The bellwether of this change has been the concept of separating "work" from "life". These used to be separate and distinct, with a hard boundary delineating the two for a majority of the time. With people spending long hours at work through the latter half of the "Third Industrial Revolution", the concept of a work-life balance grew to prominence and into our lexicon. The concept of a work-life balance has evolved to be the concept of "work-life integration" for office workers. The COVID-19 pandemic and the near two years of virtual working and virtual schooling has faded the line between office and home. Burnout of workers as a concern area was thrust into the forefront during the pandemic as workers reinvested the times they had ordinarily spent in a commute, at their desk in front of their computers. The monotony and repeatability of one video conference after another, staring at a screen for 10 to 12 hours certainly wore down workers across all age demographics. As the adage goes, there is a fine line between "working from home" and "sleeping in the office", and the pandemic challenged our collective notion of what it truly means to decouple from a screen for any length of time.

It's easy to attribute the always-on, always-connected nature of our lives today to the smartphone, and more specifically, the iPhone. However, we forget that the blurring of the boundaries between work and life did not start with the iPhone in 2007, it started in the late 1990s with Research In Motion and their groundbreaking device, the BlackBerry.

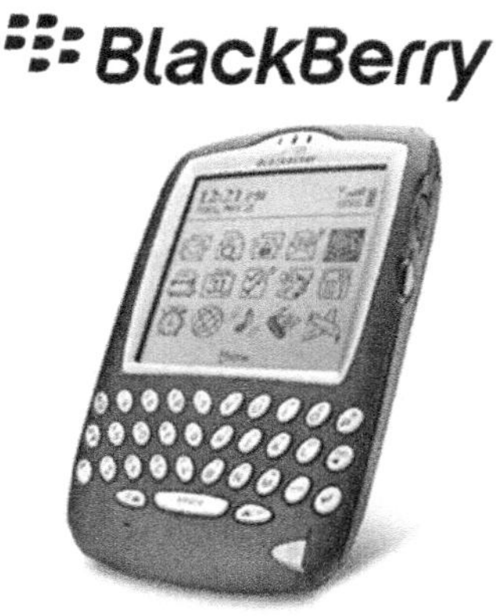

Image 12 – The BlackBerry. Image Credit: Research In Motion

Why was the BlackBerry and its distinctive QWERTY keyboard so revolutionary? It is simply because it is easy of for us to lose sight of the fact that the BlackBerry marked a massive cultural shift in our society, the impact of which cannot be underestimated. Yes, it is for sure that smartphones, perhaps starting with the iPhone in 2007, allowed us to be always connected and tethered to the internet. However, take a moment and think about life before the BlackBerry. Think about your work-life balance before the late 1990s. It is for certain that lots of people spent long hours in their corporate jobs leading up to the late 1990s, and sure, some people took their work home with them – physical files and physical folders – that some worked on at home in the evenings or on the weekends. Catching up or keeping up with work meant trudging materials to and from the office. And yet, there was little to no communication with the office (or anyone in the office). For the most part, people were truly "offline". The BlackBerry changed that with its focus on corporate email being available from anywhere that mobile cellular service was available.

The BlackBerry, in its heyday, was more indispensable to thousands of executives and office workers for this very reason – allowing people the ability to stay connected with their jobs and their coworkers. As a matter of fact, when President Barack Obama came into office and insisted on keeping his BlackBerry device, the White House had to famously "thin down" the device's functionality for national security reasons. For thousands upon thousands of professional white-collar employees, seeking to be ever connected to their work, ordinary cellphones did not simply cut it. The use of BlackBerry served as a compounding effect. If an executive got a BlackBerry, his or her leadership team, not wanting to be disconnected from their boss – and hoping to avoid coming back into the office with an overflowing email Inbox after their manager had sent them multiple emails over the evenings or weekends – followed suit and got themselves a BlackBerry as well. Then so did the employees who reported into them for the same reason that they did (to stay connected with their immediate managers). Then employee 'Bob' got himself a BlackBerry because his peer 'Sally' had gotten herself one, and Bob did not want to give their manager the impression that Sally is "more committed to work than he is, because she replies to their manager relatively quickly, regardless of whether it is in the evening or over the weekends". So addicted did office workers become to their BlackBerry devices and the keyboard that indoctrinated people to type with their thumbs, that the pejorative term "CrackBerry" was coined. What result did this compounding effect have on us as a society? Well, before the QWERTY keyboard that conditioned many people to type with their thumbs, BlackBerry devices were more ubiquitous than iPhones during the late 2000s. No one knew any better and they did not know it at the time (since there was simply no precedent), but this entire BlackBerry-addicted working generation was about to fundamentally change work-life balance as we know it.

Before there was a BlackBerry in the hands of most corporate workers, despite sometimes taking work home with you, an employee could shut down their desktop computer (and with that, access to corporate email) at the end of their workday and simply go home without being able to check email until the next workday morning. Shutting down your computer on a Friday afternoon meant that you would have 100% of a full weekend without potentially interacting with anyone in the office. You were, for the most part, fully disconnected. It's not just that there was no expectation that you had to check your email off-hours, it's that you just simply had little to no ability to do so. Think about the difference between twenty-five years ago to today. Most employees shut down their laptops, but regularly check corporate emails on their smartphones – either after dinner, during dinner, during the commute home, before bed, in bed, in the bathroom, at the grocery store, at the ball field, at the movies, and so on. Unless you are able to find the miniscule number of places where your cellular service is sporadic or Wi-Fi is unavailable, you are connected – period. And even though employers might have zero expectations that employees check email vociferously off-hours - and most employers and managers urge employees explicitly to not do so – it is more common for employees to do so than not to. It is easy to lose sight of how impactful a cultural shift this has been. And lest we need to remind ourselves yet again – none of us in any job, in any company, in any industry – could have imagined a life where the line between work and life would be so incredibly blurred.

In the following chapters, we will examine BlackBerry's demise as an illustrative case study to ensure we as employees do not replicate the same missteps that (extinct or troubled) corporations have made.

THURVIVAL: Adapting and Adjusting To The Changing Environment

Let's revisit where we began this chapter. "It is not the most intellectual of the species that survives; it is not the strongest that survives; but the species that survives is the one that is able best to adapt and adjust to the changing environment in which it finds itself" (Megginson, 1963) – Dr. Leon C. Megginson. The "boiling frog story" is a popular metaphor used in business. The metaphor describes a frog that is being gradually boiled alive. The story is based on the premise that if a live frog is dropped into a pot of boiling water, it will immediately jump out. However, if the same live frog is put into a pot of lukewarm water and if this pot is slowly brought to a boil, the frog being unable to realize the danger, will be gradually cooked to death. This metaphor is often cited in the business context in a relatively positive sense by change management professionals and leaders when an organization is seeking to introduce a new program or implement a new process. They posit that introducing a change, particularly an impactful one of some magnitude, gradually, and potentially in increments, has the best chance of success. It ensures that employees can get acclimatized to the changes, and as these change increments become "normalized" and a part of their routine, additional increments can be implemented. This mitigates employees' resistance to the new program, increasing its overall odds of success.

The very same allegory is also used as a clarion call to urge people to urgently pay attention to even the most gradual of changes, because not doing so can yield undesirable outcomes. The metaphor is used to convey that the inability to recognize seemingly gradual changes to your surrounding environment will only result in the changes increasing in complexity and severity. The lesson conveyed by this interpretation of the metaphor is that failure to react to these changes and not proactively taking measures to address them will result in certain catastrophe. The metaphor bases its core premise on a basic facet of human psychology. Human beings are generally more accepting of changes that are introduced incrementally or steadily. We accept these gradual and incremental changes even though these changes, like in the case of work email following you everywhere, seemingly are in control of us, and deliver fundamental changes to our societies and our lives. And yet, there comes a day when we awaken to realize that the impacts of these changes are an inextricable part of our lives and it is nearly impossible for us to take control of these changes and their impacts. Just as is illustrated by the fact that it is nearly impossible for us, as a society now, to go back to the days where employees could shut down their desktop computers on a Friday afternoon, and not look at their corporate email until they returned to work on a Monday morning. In the grand scheme things over the past twenty-five years, this cultural change can be considered as gradual, starting with the BlackBerry and becoming commonplace with the rise of smartphones.

To be able to "**thurvive**" in the 21st century, we need to be able to adopt and adapt faster than we have ever before. Adapting to a rapidly changing environment around us and adopting change as an omnipresent facet of 21st century professional life, we need to shift our mindset towards change. In doing so, we need to be able to change faster than we have ever before. "It is not the most intellectual of the species that survives; it is not the strongest that survives; ***but the species that survives is the one that is able best to adapt and adjust to the changing environment in which it finds itself*"** (Megginson, 1963). The time between the "lukewarm water" coming to a "full boil" in this era will be mind-numbingly quick. The metaphorical frog in this 21st century version of the allegory would not have the time to adjust itself to its environment before it is too late unless it is able to adapt and adjust rapidly. We cannot allow ourselves to be the frog that is unable to rapidly adapt and adjust to its rapidly changing environment.

Chapter Nine - Resistance Is Futile

Albert Einstein famously stated that "Intelligence is the ability to adapt to change". American author and speaker, John C. Maxwell, is attributed with the quote, "Change is inevitable. Growth is optional." George Bernard Shaw said "Progress is impossible without change, and those who cannot change their minds, cannot change anything". Sir Winston Churchill, on the topic of change, is quoted with saying "To improve is to change; to be perfect is to change often". Jack Welch said "Change before you have to". Throughout history, accomplished and inspirational figures across all walks of life, across many generations, and across vocations, have continued to espouse the value, need, and often, the urgency of change. What is it about "***change***" that makes it one of the most quoted and most quotable words in our lexicon? It is likely because despite change being the one thing you can consistently count upon, it is also the thing that can inspire the greatest excitement and anticipation, but also the greatest anxieties and trepidation. Growth and progress are only possible upon overcoming these anxieties and trepidation towards change.

The topic of change, and how humans react and respond to change has been extensively studied across every field of study. The concept of change, and how to manage it, has been a prevailing theme for deriving and applying lessons from in fields ranging from academics to literature to philosophy, and from the corporate world and to psychology. Change, and a person's reaction and response to it, is a deeply rooted, deeply personal, highly individualized, and highly subjective experience. Most corporate training programs lean into thriving through change, and leading through change as a core part of their curricula (depending on the learners these programs are geared towards, that is - individual contributor employees or leaders). In understanding that there is a famous quote, or several famous quotes that speak about change, or implore change - ranging from ancient philosophers such as Heraclitus to Confucius to Socrates to modern day corporate icons such as Jack Welch and John C. Maxwell – "Change" as a concept has been studied through time, from various perspectives, and dimensions.

Our personal "**thurvival**" in the 21st century is going to be entirely dependent on our personal ability to manage and respond to change. Change that will be around us at an enormous scale, the magnitude of which, we have yet to have experienced. Change that will seemingly accelerate from a day-to-day basis, making the task of absorbing change that has just occurred - before being faced with yet another change - to be challenging. Change that will require us to push ourselves outside our comfort zones and be comfortable with being uncomfortable. Change that will force us to "learn-unlearn-relearn" at a rapid pace so that we can position ourselves to survive and thrive amidst the "influx of the abundant flux". To understand how to "**thurvive**" and manage change in the 21st century will require us to understand how to "learn, unlearn, and relearn". To better understand what "learning, unlearning, and relearning" means as Alvin Toffler advised, we need to gain a better comprehension of the WHY behind millions of people across the world struggle with technological progress, especially when it comes to their line of work. We will also examine the technological adoption paradox - the seemingly contrary manner in which most of the same people who resist technological change at work are more than able to absorb technological progress when it comes to their own personal lives.

Why People Resist Technological Progress

Let's all take a step back for a moment and catch our collective breaths. If you leave any preconceived opinions around technology - especially technology in the workplace aside - and think about where we are, it is quite remarkable to witness where we have come from just in the past twenty-five years, and where we might be headed in the next twenty-five. An achievement for us collectively as a species; a species that believed that the

Earth was the center of the known universe just a few hundred years ago. When you do take a step back and evaluate all the conveniences of everyday life in the 21st century, thanks entirely due to technology and technological progress, you will realize what an amazing era we are currently living through. The digital age. The information age. The hallowed "Fourth Industrial Revolution". We are in an age where technology is inexorably linked to each and every facet of our existence. Our ever-increasing reliance on technology in this era is seemingly directly proportional to the exponential rate of technological advancement. The greater the advancement, the greater the inseparable reliance on technology. Technology is advancing at such an accelerating and breakneck pace that it is disorienting even the most change-hardy human beings. The faster the advancement, the greater chance of people being left behind.

Things are changing around us so fast that even the most nimble and agile individuals, and businesses, struggle with the continually evolving and continually accelerating pace of change. Why is that? Yes, we have no precedent to draw from, and it is also true that no generation/s before us have experienced this dramatic a pace of technological advancement. However, what is it that is ingrained in us - something deep rooted – that has resulted in millions of us in either falling short of being able to prepare to capitalize on the opportunities ahead, or worse, falling short on being able to see the risks ahead. Overcoming this technological change resistance is not comparable to human beings overcoming their fears about change and emerging from the safety and shelter of their cave dwellings in order to find sustenance, or inventing tools in order to make it easier to obtain this sustenance, or any other such seminal moments in human history. This era is as unique in our evolution as was the development of tools, the creation of the wheel, or taming of fire – all times in history without precedent. As we learn about our technological journey over the past several decades and into the next several decades, it is important to understand what we are up against. It is important to know fully well what it is that we are pushing back against the "active inertia" that will help us "**thurvive**" in the 21st century, and how we can overcome it. The answer boils down to the fact that we are the ones that have the ability to overcome things that are stymying our abilities to recognize and respond to the work ahead.

There are, in my view, two main reasons why people resist technological progress at work:

1. Our inability to fully grasp and process timescales outside our own lifetime.

2. The paradigm of change as relates to technology:
    a. The pace of change
    b. Being able to comprehend the change in time
    c. Resistance to change

Let's explore both these broad categories a bit further.

1. Inability to Comprehend Timescales

The first of these "double whammies" is our inability to grasp the sheer scale of time. The average human lifespan is 72.6 years according to a United Nations (UN) estimate of global average life expectancy in 2019 (Roser, Ortiz-Ospina, & Ritchie, 2019). According to the same UN report, at the beginning of the 19th century, no country in the world had a life expectancy longer than 40 years. The report goes onto start that "Globally the life expectancy increased from less than 30 years to over 72 years; after two centuries of progress we can expect to live much more than twice as long as our ancestors. And this progress was not achieved in a few places. In every world region people today can expect to live more than twice as long." (Roser, Ortiz-Ospina, & Ritchie, 2019). One of our biggest challenges to overcome when we think about conquering the 21st century and getting our heads wrapped around the pace of change, is to be able to think about timescales beyond our own lifespan and our generational perspective starting with our grandparents and ending with our grandchildren. What does that mean? It means that even though we understand the concept of time - and that we can comprehend timescales, and think about when the Second Industrial Revolution occurred in relation to our being alive - we really do not have the ability to fully appreciate time beyond two generations before us (our grandparents) and two generations after us (our grandchildren). Our perspective always operates within these parameters.

How often have we heard from our parents and our grandparents "back in my day, we did not have x, or we did y things this way."? How often will we tell our children, our grandchildren, or other younger family members the very same thing? The youth belonging to Generation Z today have a very different frame of reference than their parents and their grandparents, just as we did in relation to our parents and grandparents, and just as those who belong to Generation Z today will have in relation to their children and grandchildren. Here are a couple of facts to help better illustrate our inability to look at time in a holistic perspective beyond our own lifespan. If you belong to Generation X or are an older Millennial, and you hear the phrase "30 years ago", it is natural to think about the 1970s. At least I personally do. It's jarring to think that, in fact, 30 years ago in relation to 2022 was 1992. While we're thinking of the late 20th century, 1980 is as far apart to 2022 as 1980 is to 1938. As an example, thinking back to 1980 from 2022, would be akin to a person in 1980 thinking back to 1938 – a year before World War II started, and a year when that marked the appearance of the world's first superhero (Superman in Action Comics, June 1938). People who call themselves "children of the 80s" are 40+ year old adults as of 2022.

Our inability to grasp but not fully understand timescales in relation to own lifetime leads to us losing perspective. Earlier in this book, we talked about the fact that most of us sent our first ever email less than twenty-five years ago. Most of us sent and received our first-ever text message less than twenty-five years ago. We hailed our first Uber ride less than ten years ago, and swiped our first digital boarding pass to board an airplane less than ten years ago. Until these facts are thrust in our face, we tend to ignore things that are directly under our noses. If we can tend to forget and take for granted these advancements that changed our lives less than twenty-five years ago, how can we possibly comprehend how dramatically life has been changing in the tail-end of the Second Industrial Revolution? How can we get our heads wrapped around the fact that less than 100 years ago, less than half of the United States had access to electricity to power their homes and extending onto that, none of the things that we use today – consumer electronics – existed because there was no means to power them.

Our inability to see this bigger picture perspective limits our ability to experience the changes happening around us in real-time, because we cannot adequately process changes that have happened just preceding our immediate perspective. While delving into the human psyche is not the focus of this book, in order to demonstrate how distorted our relative sense of time is, it is illustrative to explore some examples, courtesy a compilation that appeared in a BuzzFeed article in 2016 (Golder, 2016). If nothing else, it is a fun trip through history to draw attention to our inability to quantify, yet not fully comprehend timescales.

1. In a prior chapter, we cited an example of companies in 2021 looking towards CRISPR to genetically resurrect the extinct Wooly Mammoth (Neuman, 2021). For timescale illustration, note that the Wooly Mammoths were still alive when the Great Pyramid at Giza in Egypt was constructed. As a matter of fact, the Woolly Mammoths went completely extinct around 1560 BCE (Wilkins, 2012), a whole 1000 years after the Great Pyramid had existed.

2. Speaking of the Great Pyramid of Giza, the Great Pyramid was older to the Romans than the Roman Empire (Britannica T. E., n.d.) is to us in 2022.

3. Continuing to explore Egypt in timescales, Queen Cleopatra lived closer to the invention of Snapchat than she did to the construction of the Great Pyramid at Giza. Cleopatra died in 30 BCE, slightly over 2000 years ago, while the Great Pyramid was completed around 2560 BCE, a difference of more than 2500 years.

4. Restoration of an entire extinct species such as the Wooly Mammoth is the plotline to the Jurassic Park series (IMDB, IMDB All Jurassic Park Movies, 2017). The series depicts multiple species of dinosaurs cohabiting in the wild after being restored like the real world efforts with the Wooly Mammoth. Two of the more memorable dinosaurs through the series are the Stegosaurus and the Tyrannosaurus Rex, but in relative terms, the Stegosaurus was older to the Tyrannosaurus Rex than the Tyrannosaurus Rex is to humans in 2022. The Tyrannosaurus Rex emerged as a species 86 to 65 million years ago, a whopping 74 million years after the Stegosaurus went extinct.

5. George Washington, the first President of the United States would have never known about dinosaurs of any type. President Washington died in 1799, approximately 40 years before the existence of dinosaurs in our planet's past was even discovered.

6. Speaking of US Presidents, the United States has had 18 presidents as of 2022 for as long as Queen Elizabeth II had been alive.

7. The United States is a relatively young country. In fact, those born in 1969, have witnessed 20% of US history (about a fifth of the age of the United States).

8. Pivoting to technology, let's examine an example of a popular corporate and consumer electronic device (we will explore consumer electronics and their rapid evolution further later in this book), the fax machine. People

were still traveling the Oregon Trail (Editors, 2009) when the fax machine was invented. The first notable wagon train that comprised the Oregon Trail commenced journey in 1841 and an early prototype of the modern fax machine was first developed by Scottish inventor Alexander Bain (Encyclopedia, n.d.) in 1843.

9. Tying together US Presidents and the fax machine: Samurais as a caste were officially abolished in Japan in 1867. This was 24 years after the fax machine was invented in 1843. President Abraham Lincoln was assassinated at the Ford Theater in 1865. So theoretically, a samurai could have sent President Lincoln a fax between 1843 and 1865.

10. And finally, as a segue into the accelerating nature of technology in the 21st century, consider that humans made it to the moon and back in 1969, just 66 years after the Wright Brothers invented human flight in 1903. In less than one human lifetime, humans developed the technology to take flight and successfully completed a round-trip mission to our closest celestial neighbor.

As renowned astronomer Carl Sagan outlined in his 1980 television show, Cosmos: A Personal Voyage (IMDB, Cosmos , n.d.), if you compressed the entire history of everything, going back to the Big Bang, into a calendar year, all of human history would only take up an insignificantly small amount of time. As Carl Sagan explains:

"Imagine that the age of the universe, 13.82 billion years, is compressed into only one year. On this Cosmic Calendar, the Big Bang happens on January 1st at midnight, and we are at the end of the year, midnight on December 31st. Now look backwards: Down here, the first humans made their debut around 10:30 p.m. on December 31st. And with the passing of every cosmic minute - each minute 30,000 years long - we began the arduous journey towards understanding where we live and who we are.

11:46 PM - only 14 minutes ago, humans have tamed fire.
11:59:20 PM - the evening of the last day of the cosmic year - the 11th hour, the 59th minute, the 20th second - the domestication of plants and animals began, an application of the human talent for making tools.
11:59:35 PM - settled agricultural communities evolved into the first cities.

We humans appear on the cosmic calendar so recently that our recorded history occupies only the last few seconds of the last minute of December 31st. In the vast ocean of time which this calendar represents, all our memories are confined to this small square. Every person we've ever heard of lived somewhere in there. All those kings and battles, migrations and inventions, wars and loves. Everything in the history books happens here, in the last 10 seconds of the cosmic calendar." (TKSST, 2012)

Not only are we disadvantaged because we are unable to readily comprehend timescales, but this problem is exacerbated by the fact that the pace of change has been so incredibly rapid that we have been unable to wrap our heads around any one single major cultural and societal change before a dozen other such dramatic changes – both personal and professional – wash over us like a tidal wave, a metaphorical tsunami of changes. This is what millions of workers experience towards the end of one industrial revolution, and the transition into the start of the next one. A tsunami of changes that are well outside any particular individual's control. While we struggle to ground ourselves to where we are at any specific point in time - with our inability to grasp timescales as a backdrop - millions are left to grapple with the massive disruption that a transitionary period between industrial revolutions brings. The technology fueled changes of the "Fourth Industrial Revolution" are likely going to be the largest of these metaphorical tidal waves, and as of 2022, this technology tidal wave is only just getting started.

2. Change

Through all of our inability to fully comprehend timescales beyond our limited purview, the basic underpinning that most of us are naturally wary of change at a most basic level, still remains. Some of us are enthusiastic proponents of change, some of us are eager champions of change, and yet, at the most basic, we can be change-averse creatures. Change triggers uncertainty, and uncertainty fosters a sense of fear of the unknown. We value consistency and predictability. Consistency and predictability, in our "lizard brain" - characterized as "the part of our brain" that is responsible for our most primitive or that our most self-interested behavior is attributed - the job of which is simply to protect us against threats by letting our instincts protect us against threats.

Our brains aren't hard-wired to make us innovative and adventurous, they are hard-wired to keep us alive. This "lizard brain" or "reptilian brain", within the context of external change, translates into our ability to manage and control our environment in order to keep us safe and secure against all manner of external threats. Generally speaking, in a group of ten people, a third of this group will be enthusiastic proponents of change (known as "promoters"), a third are passive (known as "neutral"), and can sway towards being enthusiastic proponents, or at the least, willing participants of the change, and a third are resistors (known as "detractors"). This latter group of resistors not only resist any change, but might seek to actively sabotage it (like the Luddites). In any organization that is seeking to introduce change, the focus should always be on tamping down potential toxicity emanating from the resistor group, and converting the "passive thirds" into willing participants - if not downright champions - to promote change. However, no matter how enthusiastic a promoter group is, even though they might have completely bought into the change, and would actively promote it within their companies, somewhere deep down inside, each of these promoters naturally harbor a certain amount of trepidation about the change – the fear of the unknown. They are able to overcome their deepest fears, their "lizard brain", to actively champion change and change themselves in order to rise and meet this change. These promoters then, in turn, influence the individuals in the passive group to join them in being a part of the overall change.

Our goal as we enter the "Fourth Industrial Revolution" is to be promoters – active champions and influencers - of our own careers, adopting and adapting to the disruptive changes that abound all around us. To be effective champions, we will need to survive some changes, and thrive amongst others. To be best equipped to be able to do that, we will need to understand the concept of technological change and our response to it a bit further, something we will do in the subsequent chapter.

Chapter Ten – Change: The Pace of Change

Those who are averse to change are averse to it because they tend to focus on the disruption the change can bring, rather than focus on the desired end result of the change. They tend to fixate on what they would potentially be losing versus what they could potentially be gaining. Some deny that the change is happening, others deny that change will happen to them, choosing to believe that someone else, somewhere else, will feel the impact of change.

By now, we've firmly established that the pace of technological change in the first quarter of the 21st century has been absolutely breakneck, and it is only continually accelerating. We've established that there aren't very many facets of our personal and professional lives that will remain untouched by this accelerated and accelerating change. We have spent some time in establishing that just the past twenty-five years should serve as a good gauge of what the next twenty-five years will look like. No matter how much we believe we know we are headed, technologically speaking, we should be prepared to understand that the reality of change will likely exceed the boundaries of our speculative imagination. In the preceding chapter, we discussed that there are two main reasons why people resist technological progress at work, the first being our inability to fully grasp and process timescales outside our own lifetime. The second reason is the paradigm of change as relates to technology; specifically: a. the pace of change, b. being able to understand the change in time, and c. the resistance to change.

Change, The True Constant

We are so busy in living our daily lives and going about our packed schedules of the early 21st century, that we can become desensitized to the accelerating nature of technological changes around us. We take for granted that technology will progress and society will adapt this progress with it. However, there is a distinct dichotomy between how we, as individuals, embrace technological progress versus how we, as workers, seemingly do. The collective nature of our slow adoption of technological progress within the workplace aggregates to resulting in entire industries, or sectors within industries, being laggards in digitization. We need to recognize that there is a palpable disconnect between how we have reacted to technologically change in our personal lives, versus technological change in our professional lives. This recognition is important for us to be able to apply the same mindset towards technology that we do as consumers towards our mindset as workers. Why have we been adept at keeping up with technological change as consumers when the same individuals have not been as enthusiastic adopters of technological change in the workplace? The answer simply boils down to adoption of change as a willing personal choice in the case of us as a consumers, versus adoption of change thrust onto us within our workplace - with workers having little to no choice in the matter.

It is dismissively stated that x industry is slow to adopt digitization, or y sector of the industry is a laggard in contrast to others when it comes to technology. But to characterize an industry as slow to adopt technology distracts from the point that industries and sectors are made up of companies, and companies are made up of people. The same people, who, as consumers will annually upgrade to the latest and greatest iPhone, but will be averse to continually learning and relearning new tools and skills to keep their companies, sectors, and industries vibrant, and on track with technological change. This is the technology adoption paradox. We as consumers are more than comfortable with technological change, but can, in aggregate, slow the march to technological vibrancy in our organizations.

In this "Fourth Industrial Revolution", it will become imperative for each of us to apply this consumer mindset to technological adoption to our workplaces and to our careers. Our personal lives will be enriched by continued advancements in technology in the next twenty-five years, and just as we have in the past twenty-five years – we will continue to embrace this progress and these changes in our personal lives. Our professional lives on the other hand, might continue to potentially be laggards - just as the past twenty-five years have illustrated. Here are a few illustrative examples of the dichotomy between our progress in adopting technology as consumers, versus demonstrating a comparable level of progress within our workplaces.

Color TV

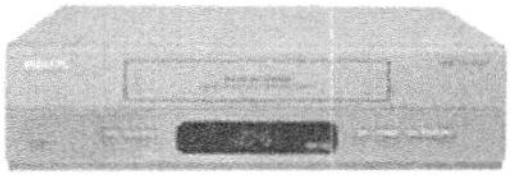

VCR

Camcorder

Portable CD Player

Portable Gaming

Gaming Console

Image 13 – Most Popular Consumer Electronics in 1998

Image 13 above depicts the most popular consumer electronics in 1998, less than twenty-five years ago. Starting with the top left of the image, color TVs, especially portable color TVs were highly popular across the United States because TVs became sharper, more portable, and most of all – cheaper. The cost effectiveness of televisions meant that most households did not need to have one giant, immovable, heavy television in their living room or their den that every person in the household needed to cluster around, typically all having to watch the same channels and shows. These portable, color TVs came at a price point that allowed households with even the most modest of incomes to purchase multiple televisions in a home. From teenagers being able to watch their own shows such as "Dawson's Creek" (1998) and "Buffy the Vampire Slayer" (1997), while their dad watched the news or sports in different part of the house, to college students having a TV set in their own dorm room and connect them to video gaming systems such as the Nintendo 64, TVs became as pervasive in every room in the house as did furniture that belonged in that particular room. The TV is also a good example of a technology that experienced explosive growth. In the 1950s, roughly 8% of American households had a television set. In the 1960s, in just a decade, 89% of American households owned a television set.

The popularity and widespread availability of portable color TVs, led to the consumer electronics explosion of ancillary devices and peripherals to augment the TV experience. These peripherals included the video cassette recorder, the VCR. The VCR is an underrated device when you consider the impact that it has had in serving as a catalyst that transformed home entertainment. For the first time, people were able to purchase or rent a wide range of entertainment, from movies that "came out on tape" a certain period after they left the movie theaters, to sporting events, television shows, etc. Imagine how much more profitable the movie and television industry became after they could continue to deliver movies, TV shows, and sporting events, after these had completed their run in movie theaters or were done their episodes on TV channels. People collected VCR tapes and started to develop quite a rich collection of TVs and movies in their home video libraries. Not only did VCRs deliver untold profits to studios and corporations and reshape home entertainment as we know it, it also spawned an entire, highly-profitable video rental industry. Blockbuster, Hollywood Video, Movie Gallery, and other video rental companies, as well as stores that sold videos such as FYE, Sam Goody, Virgin Megastore, Borders, were highly profitable businesses. Another peripheral device that consumers connected to their TVs that exploded in popularity was the Camcorder. Camcorders got smaller, and much more portable through the early 1990s, allowing families to capture videos of their kids special occasions, birthdays, sporting events, weddings, vacations, and so on, allowing them to memorialize these events more than what the static imagery of photos would allow. Connecting them to home TV sets, and family vacation videos could now repeatedly be shared with others (willing viewers or otherwise). You could record a family event, make multiple copies, and distribute the same memory to multiple members of your family or participants of the event, and everyone could share in these memories.

Although the Compact Disc format, CDs, had existed since the 1980s, CDs became the default media for storing and playing music through the latter half of the 1990s. CDs could hold a lot more music than cassette tapes, with a significantly superior audio quality to cassettes. Music CDs and cassette tapes coexisted as music media playback devices for several years, before CDs became the dominant format to store, distribute, and play music. People could "burn" music from onto discs using rewritable discs, and the era of "mix tapes" gave way to music CD mixes where each individual could create a customized playlist. Just as the portable cassette tape player, such as the Sony Walkman® was transformational to music that people could carry with them anywhere and everywhere, the Sony Discman® and portable CD players had the same effect for transporting an order of magnitude more music than tapes would allow for. CDs were undeniably big business, with stores such as Tower Records, Virgin

Megastore, Sam Goody, etc. having earned significant revenues through the heyday of this media type. Portable CD players promised better music quality, the ability to store and playback a greater quantity of music, and "skip resistant" media.

The popularity of video gaming began in earnest in the 1980s with consoles such as the Atari having become household names. With modest graphics and limited number of games, the evolution of these platforms resulted in the arrival of modern (for the time) consoles such as the Nintendo 64 in 1996 (named for its 64-bit computer processor that delivered better graphics and 3D image capabilities), and its main competitors the Sony PlayStation, and the Sega Dreamcast (for a limited time). Portability gained popularity, as evidenced by the portable CD player, the portable TV, and the rise of cellular phones (not smartphones). In keeping, portable video gaming systems, that had been around since the 1980s became much more sophisticated with platforms such as the Nintendo Gameboy in color.

Now let's take a look at some of the most popular business and personal consumer electronics in 1998. Image 14 below depicts some additional most popular consumer electronics less than twenty-five years ago. What is interesting about this set of products is that they had use within our workplace and our homes.

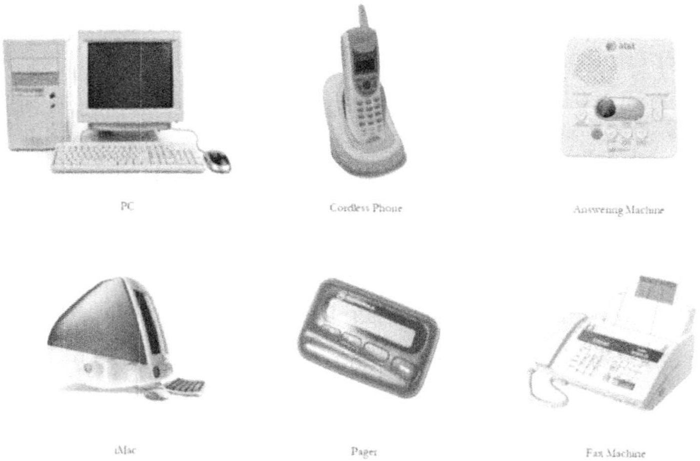

Image 14 – Most Popular Personal Consumer AND Business Electronics in 1998

Whether you had a PC or an iMac in your house, most homes had a dedicated "computer area" in the house where the computer sat. People within a household took turns in using the computer. It was especially treacherous when home internet via dial-up modem first came about. Many-an internet surfing session, achieved with great difficulty via "dialing up" to the internet using America Online (AOL) or any such service providers, was interrupted in an instant by someone in the house answering an incoming phone call. Phone and internet services used the same exact "wire" - and "dial up" literally meant dialing into these service providers via a 28.8K, and later, 56K modems, and subjecting oneself to the familiar, telltale screeching sounds of "dialing into" the world wide web. Windows 95 revolutionized personal computing and took great strides into making the vision of "a PC in every home" a reality. Before long, families were using PCs for everything from college essays using Word Processors to household budgeting via spreadsheets, to seeking individuals with like-minded interests on the internet via "usenet" groups (rudimentary online forums). In the age of the dawn of the PCs, retailers such as Circuit City, Radio Shack, and Best Buy flourished. Of the three aforementioned retailers, only Best Buy continues as a vibrant enterprise today, with a majority of their sales emanating from consumer electronics such as televisions and computers.

The pager, also referred to as a beeper, was quite a handy gizmo to "page" individuals who didn't yet have a cellular telephone. No matter where you were, a pager could send a limited character message notifying you to contact whoever was attempting to reach you. This was a really handy tool for doctors and physicians when hospitals needed to reach them during patient emergencies or when urgent matters warranted it. The term "pager" itself borrows from hospital lingo when a doctor is being summoned "paging Dr. Jane Doe". Although they've been around since the 1950s/1960s, pagers grew to prominence through the 1980s and into the 1990s, before the pervasiveness of mobile phones rendered them redundant, and on the cusp of being irrelevant. One might argue that modern-day text messaging, "Simple Messaging Service" (SMS), in its simplest textual format, is an evolution to the one-way communication that pagers used to provide. I personally had a pager, given to me by my employer, as a requirement of my first job out of college. Being the lowest man on the team's "totem pole" at that time, I was responsible to carry a pager so that I could receive a page in the middle of the night that a software assembly had failed and that I had to take urgent action to restore service. The year was 1999, and although the pager inspired in me a misplaced sense of self-importance, a device to clip to my belt and live out my Batman utility belt aspirations, it never actually buzzed for work reasons for the two years I had possession of it. Eventually, I turned in the pager for two reasons: 1. I was not the lowest guy on the corporate ladder, and 2. Mobile phones rendered pagers irrelevant.

Through the late 90s, even as voicemail on mobile phones (and pagers) emerged as viable alternatives to answering machines, these devices were a common accessory to home phones across most households. Not a new invention per se, the world's first answering machine was invented by Swiss inventor Willy Müller in 1935. The three-foot-tall, complex machine did not lend itself to be particularly portable, but found itself to be popular with the Orthodox Jewish community since they were forbidden to answer the phone on the Sabbath. The popular home answering machine was a descendent of the Ansafone, the first answering machine sold in the United States, created by inventor Dr. Kazuo Hashimoto for Phonetel in the 1960s. By the 1990s, answering machines that featured the standard playback of "Please leave your message after the beep" - or allowed people to record their own greetings - were cheap and widely available, fueling these devices to becoming one of the most popular consumer electronics products of this decade. The answering machine harkens back to an era before SMS text messaging, where phone calls actually take a backseat to texting others. Having been a near-common presence, as home phones and landlines decreased precipitously in favor of mobile phones, the answering machine, a constant companion to the home phone through the 1990s, also has undergone a similar steep decline in being an omnipresent gadget in people's homes. Ironically, not only has the popularity of answering machines expectedly declined, reflective of the decline of landline phones, but voice calling seems to be taking a backseat of text messaging and other messaging services. Voicemail is to mobile phones what answering machines is to landlines, and today, even when people call others, they typically do so expecting to get their voicemail versus actually wanting to talk with someone. A generational gap has also emerged as younger Millennials seem to lean towards video calling as a means to communication, whereas Generation Z do so via newer social media platforms such as Snapchat and TikTok as a supplement to text messaging.

The cordless phone was quite a revolutionary creation for most homeowners and businesses alike. On the home front, most individuals over the age of 30 would remember a time of landline home phones. These telephones, especially those mounted in people's kitchens, came equipped with extra-long telephone cords. While these long cords allowed for people to walk about their kitchens, while being on the phone, they got entrapped on everything. Navigating across a kitchen, while on the phone, proved quote challenging as these long cords and the cords became entangled in people, pets, knocked over plates and place settings. The cordless phone was the panacea to people who felt tethered to a limited area by their phones. Cordless phones offered the ability to pace around the house, answer the phone in one room and move across the house (as long as you were still in proximity to the phone cradle/base). The late 1990s witnessed the presence of multiple cordless phones across most rooms in a home. This is also the period where accessing the internet required use of the same telephone line, and most of Generation Z today do not know why or in what context the term "hang up the phone" originated. As a matter of fact, when this generation gestures that they are on a call, they lay their palm down completely flat against their ear, representing a smartphone design. Just a generation prior, the gesture of being on the phone is having the thumb and pinky fingers extended out in either direction, with the three fingers in the center curled up against the palm – akin to a Hawaiian "shaka signal" / surfer "hang ten" or "hang loose" hand signal held up to the ear. Also closely associated to phones and landlines, given that a telephone exchange was a necessity for it to work, the fax machine was a fixture in most businesses, as well as in some households. The fax machine was a popular consumer electronic product of the late 1990s – albeit one that is still in use today, particularly in industries that have been slow to adopting technology.

The Pace of Change in Less Than Twenty Five Years

Now, here is the kicker. Everything that we've reviewed in the preceding section across personal consumer electronics as well as those that have had applicability across the business world – each and every one of these products – have either been rendered obsolete, or made exponentially better and collapsed into a single device that almost each and every one of us own – and transport in our pockets or our pocketbooks.

Image 15 – The iPhone (iPhone 11), credit Apple

Smartphones, such as the iPhone pictured above, or an Android-based device, have encapsulated the features and functionality of every most popular consumer electronics products in the late 1990s, and made these devices relics of a bygone era. What is truly remarkable is that this "bygone era" is less than twenty-five years ago, just barely above what is the legal age to consume alcoholic beverages in the United States. And once again, it raises the question – with this breakneck pace of change, what products and services that we rely on today will be rendered obsolete in the next twenty-five years?

The Romans could stake a claim to transformative advancements with construction of a network of roads to the constructing the aqueducts. It could be said that the industrial age has witnessed revolutionary changes with the advent of electricity, heavy machinery, and automation, to advancements in transportation from the steamship, to railroads, the automobile, and air travel. Every generation before us has undergone and witnessed some level of change and progress, and every generation might have considered a few innovations of their time as true "game changers". If change has been constant, what is it about the speed and pace of the change experienced by our living generations that stands apart from anything and everything before it? Why is it that this age we are living through simply feels different than any other? It feels different because it is, and vastly so. This era appears and feels to be vastly different than every other era before it because we, the human inhabitants of this increasing digital planet, are changing much slower than the technology around us.

Chapter Eleven – Change: The Resistance to Change

Change resistance is one of the key inhibitors to "**thurvival**". Humans are creatures of habit, especially when these predictable, repeated habits provide us with structure, a sense of comfort, and above all – safety. It is this aspect of safety that has been hardwired into us for thousands of years. This is why having humans leave the safety of their caves to explore, find sustenance, and expand into new lands was such a remarkable feat. This act of overcoming their fears, and perhaps out of desperation to hunt and forage, allowed humanity to explore and settle (nearly) every continent on this planet – and most likely, another planet, Mars, not too long from now. It could be argued that change has been the one true constant across most of human history. It could also be argued that this constant change has also been met with constant resistance when it was first introduced.

We discussed the "Luddites" of the 18th century and the "Neo Luddites" and "Technology Resistors" of the 21st century, but technological progress isn't always purely digital in nature. Human beings have a long standing tradition of putting up resistance to the most basic of innovations and progress that we take for granted today. For instance, the humble fork encountered resistance when it was first introduced in the Europe. According to Sarah Coffin, Curator and Head of Product Design and Decorative Arts at the Cooper Hewittt Smithsonian Design Museum, "Forks were used as cutlery in the Byzantine Empire and made their way west during the eleventh century when a Byzantine princess married the doge of Venice. She brought a two-tined gold fork with her to eat fruits preserved in syrup and sugar, known as suckets. Although this may simply seem like a logical way to eat a sticky sweetmeat, the church considered it scandalous." (Eplett, 2015). Coffin explains, "The fork was associated with the life of luxury and sweets. This somehow got translated by the church as being a negative form of decadence. It wasn't associated with Christian values on the grounds that it wasn't essential to life. Instead, it was perceived as something that would be used by a seductress of the East." (Eplett, 2015).

We don't have to look too far into the past to see how a global industry that dated back thousands of years was suddenly usurped. We can simply take a look at the early 20th century when Henry Ford transformed personal transportation by the mass production of the world's first mass produced automobile, the Ford Model T. Regarded as the world's first affordable automobile, the Model T disrupted the equine-based transportation industry. Considered to be revolutionary and disruptive at the time, Henry Ford is famously quoted for stating - "If I had asked people what they wanted, they would have said faster horses." Unsurprisingly, the rise of the automobile was also met with resistance from horse breeders, stagecoach manufacturers, and all the trades that this industry supported. Horace Rackham, Henry Ford's lawyer, was advised to not invest in the Ford Motor Company by George Peck, the President of the Michigan Savings Bank in 1903. Mr. Peck is quoted as advising Mr. Rackham - "The horse is here to stay but the automobile is only a novelty – a fad." The horse carriage industry was undoubtedly a globally established business that was threatened by the seeming overnight popularity of "horseless carriages". Large, established horse carriage manufacturers at the turn of the 20th century faced a difficult choice – transform their companies to manufacturing other things, including potentially parts for automobiles, or dig in their heels and seek to win a battle that had already been lost. Some that found themselves being intransigent against change actually spent time in lobbying the public against automobiles – time that could have been better spent in reinventing themselves. Ed Klein is a great example, having posted an advertisement in the Lawrence Daily Journal-World Newspaper (Lawrence, Kansas, United States of America) on September 21st 1915, trying to convince people that horses were a better mode of transport than automobiles as depicted in Image 19 below. Highlighting that automobiles would be less cost-effective to own and maintain, this advertisement read:

"Come in and get a new harness instead of a new car and remember that Dobbin will take you through snow and mud as well as on good roads and that his carburetor is never out of order."

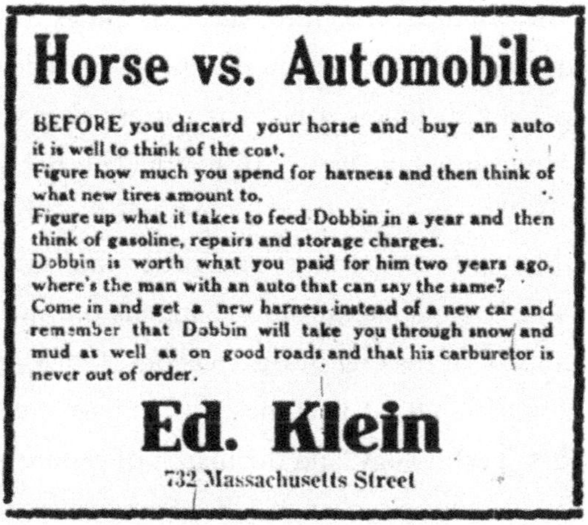

Image 19: Ed. Klein advertisement in the Lawrence Journal-World (Klein, 1915) (Credit: Lawrence Journal-World)

Examining history, we can clearly see that people have always resisted technological progress and change. Despite the disruption and disintermediation that accompanied these technological changes, people in every era prior to the one we are heading into have had a lot more leeway in absorbing, adapting, and adopting these changes. Some changes progressed much more rapidly than others, but by and large, the pace of these changes historically was stretched out over a period of time versus the hyper-compressed time period we find ourselves in. In my opinion, there are two major reasons why humans exhibit technologically change resistance:

1. Comfort and Complacency - In 1946, Darryl Zanuck, an executive at 20th Century Fox, in response to the rise of television, said "Television won't be able to hold on to any market it captures after the first six months. People

will soon get tired of staring at a plywood box every night." It was clear that Mr. Zanuck had become comfortable in the idea of entertainment only being delivered via the silver screen. Humans generally favor being comfortable, comfortable in our routines, comfortable in our patterns, and comfortable in the predictive repeatability of things that are familiar. Familiarity breeds comfort because it produces known outcomes. It is exponentially less risky to follow a familiar pattern that we have become adept at versus trying something new or different. This behavior is apparent in everything we do. We have the same morning routine, the same shopping habits, frequent the same restaurants, and order the same meals. Trying something new – or being adventurous – happens on our own terms and when we are in control of making those decisions. Once you get used to a certain way of doing things, it becomes difficult to break out of these ingrained behaviors. Technology, and the march of progress however cares little about our feelings towards change. Change management is therefore a key concept when organizations rollout new platforms or technologies within their companies. Comfort isn't always necessarily a bad thing – it keeps us safe and productive (the longer you do the same thing, you get quite proficient at it). However, comfort also fosters complacency. Complacency is highly undesirable, even though it allows people to become comfortable at their tasks and use of existing technology. Complacency stifles innovation and hardens peoples' attitudes towards change. It is this complacency that compounds and results in organizational inaction in the midst of rapid transformation. It can be the death knell of an organization or an entire industry.

2. Loss of Control – In 1876, Sir William Preece, Chief Engineer at the British Post Office, in response to the rise of the telephone, was quoted as saying, "The Americans have need of the telephone, but we do not. We have plenty of messenger boys." Rather than focusing on growing through change and due to change, people have historically and instinctively adopted a "change is optional" mindset when it came to the introduction of new technology – especially within their trades and professions. Although some

resistance to change has historically been across every facet of life, only in the past twenty-five years have we witnessed people embracing technological progress in their personal lives, but continuing to resist this progress in their professional lives. There is a deceivingly gradual technological pace of change within our homes and at our workplaces, between our personal technology progressiveness and in our professional settings. Most employees in white collar jobs can readily rattle off a dozen archaic technologies and platforms that their organizations use to this day. However, follow them home, and it is quite likely that most of their consumer electronics do not date back more than five years on average. Regardless of age group and generation these individuals belong to, most rely on multiple apps on their smartphones to live their daily lives. From using rideshare services such as Uber to renting accommodations via Airbnb or ordering a variety of goods via Amazon, we as consumers are far more technologically progressive than we are as professionals – regardless of the actual profession we serve in.

The simple reason for this dichotomy is one of choice and control. We are - for the most part - firmly in control of our technological choices in our personal lives. We choose to upgrade to the latest iPhone model. We choose to purchase products online via Amazon versus heading down to the neighborhood store. We choose to conduct our banking fully online via our bank's app, rather than step into our local bank branch. We choose, period. We are fully in control of our choices – and even though there are early adopters and laggards, defined by who amongst us embrace new technology and when they do so – even laggards at some point take the plunge and go along for the ride. In the case of the laggards, it might be due to the fact that the way they used to do things simply cease to exist because of technology progress. In the case of the early adopters, it might be their drive for "playing with the latest toys", or a penchant for showing other people that they're ahead of the pack. For those who form the vast majority of personal technology-adopting public, their rationale ranges from wanting to "keep up with the Joneses" to having their kids force them into it "because the neighbors kids have x technology", etc. In any case, we as consumers more often than not, always retain the choice. When we retain control and have choice, we tend to focus on what we gain versus what we lose. "By using Amazon, I won't need to leave my house and save x time", "by upgrading to the latest iPhone, I will be able to take higher quality pictures of my family", "by using Airbnb, I will get to experience what it is like to live like a local and probably save some money", etc. are the types of things we tend to fixate on when we choose to capitalize on technological progress.

Now contrast this with having to use a new platform or some new technology at work. New technology at work influences your job on a spectrum that ranges from minor changes to a complete and absolute disruption of the nature of your work. There are some technologies that dramatically alter a job, forcing the incumbents to learn new skills or seek to augment their professions to cohabit with this new technology. Other technologies simply seek to lucidly integrate into your jobs – attempting to make things easier, faster, smoother, or just more accurate and reliable. Some technologies are welcomed since they automate some of the rote, operational aspects of your job that you enjoyed much less than all the other aspects.

However, regardless of how intrusive a new technology is within your profession and regardless of how beneficial its influence on your day, more often than not, the selection and implementation of this technology was not your choice. Not only was it not your choice, it likely requires something from you – to learn a new platform, to step outside of your comfort zone, to do your job that you've been faithfully doing for *n* number of years differently than ever before, to educate yourself on a new business process and a new way of doing things – on top of your existing responsibilities and obligations. More often than not, the introduction of a new technology – no matter how beneficial – is a "job plus" for you. More often than not, you had little to no say in the implementation of this new technology. You had no control. You had no choice. You will, going forward, have no choice in the utilization of this new technology. You will use it, and you will like it, learn to like it, or at the least – pretend to like it to keep your gainful employment. And all of this presumes that your job function or role is still unimpeachable after implementation of this new technology – that this new technology hasn't rendered your role or function moot or obsolete.

Think about all the technology advancements that have happened around you within your workplace. Now think about all the technology you have chosen to purchase in your home and in your personal life. Which of these two "feel" like they've advanced at a much faster rate? Even if your company's implementation of new technology has felt just as fast as your adoption of technology in your personal life, is this adoption equitable across your company? There is a deceivingly gradual and dichotomy between the rate of technological advancement in your personal and professional life - deceptive because while it may seem as though you are a lot more advanced at home than at work, it might actually be that both home and work have attempted to progress at the same pace, but this has been stymied by how quickly this progress has been equitably adopted. According to an article that appeared in the Washington Post summarizing the book, "Innovation and Its Enemies: Why People Resist New Technologies" (Juma, 2016) by Harvard University's Kennedy School of Government professor Calestous Juma, author Steven Overly opens by stating "Humans have a habit of stalling their own progress. From coffee to mechanical refrigeration to genetically altered food, history is littered with innovations that sparked resistance before becoming fixtures in everyday life." (Overly, 2016).

Mr. Overly, continues, "In hindsight, opposition to innovations such as mechanical farm equipment or recorded music may seem ludicrous. But the past 600 years of human history help explain why humans often oppose new technologies and why that pattern of opposition continues to this day."

(Overly, 2016). In his book, Mr. Juma explores the human predisposition in opposing technological advancements and states that people do not fear technology simply because it is something new and different. Instead, Juma posits that people fear innovation since it can mean a loss of their identity or their lifestyle. Innovation, Juma argues, can also separate people from nature, or their sense of purpose, both of which are essential parts of the human experience.

Mr. Overly cites eight key takeaways (Overly, 2016) from his dialog with Mr. Juma regarding his book on the propensity for humans to be averse to technological innovation. These eight takeaways are enumerated below:

1. People sometimes oppose innovation even when it seems to be in their best interest.
2. Technologies that are vastly superior to their predecessors, or don't have any predecessors, are more easily adopted.
3. Resistance to new technologies comes from three key constituents, including the average consumer.
4. Humans make decisions about new innovations with their gut rather than evidence.
5. People flock to technologies that make them more autonomous and mobile.
6. People typically don't fear new technology, they fear the loss it will bring.
7. Technologists often don't think about the impact their inventions have on society.
8. Innovation is not slow, linear or incremental - but the government doesn't realize that.

The Urgent Need to Prepare for Thurvival

We have explored what our likely future in the next twenty-five years might look like in the first section of this book. We have sought to orient ourselves on the lead up to where we are today, examining the rapid and accelerating pace of technological transformation and the "Fourth Industrial Revolution" that we are at the starting line of in the second section. The third section delved into the difficulties we face in being adequately prepared for what's next for our personal lives and what awaits us in the next twenty-five years of our professional lives. Through the first three sections of this book, we can clearly see that we are facing extraordinary times ahead of us, an era without precedence. Not only would we benefit from additional preparation to fully thrive in this era, but some of us might actually be unprepared and underprepared to have our professional lives survive amidst accelerating technological change that is progressing at an exponential rate. We could be at a dangerous precipice, challenged by our own hard-wired tendencies that prevent us from adopting and adapting rapidly, the resistance to technological change ingrained into our psyche.

For companies and entire industries to survive and thrive, we - each of us – need to rewire our own thinking. We need to now, more than ever, be able to adopt and adapt rapidly. If you embody this thinking, you can thrive. If you embody this thinking, and others don't, you still need to find a way to survive. If you are able to influence others to share in your mindset, you can mutually thrive. If every single person in any organization embodies this thinking, the company can collectively thrive. The next section will give you some practical and pragmatic methods to survive and thrive, to **thurvive**.

Chapter Twelve – Change: The Past Should Not Be Prologue

The breakneck and accelerating pace of technological change allows us to exert control over everything – except the breakneck and accelerating pace of technological change.

Technology is changing very fast, making it more likely that people will be very slow to adopt to these changes. Technology is changing as fast as it is because of a phenomenon known as "accelerating change". Accelerating change has been defined as the exponential nature of the rate of technological change throughout history, with dramatic observations through recent history. Accelerating change portends ever-accelerating and impactful technological changes in the future. These technological changes might likely be the cause of lasting and profound cultural and social changes.

According to Ray Kurzweil, responding to what exponential change will mean for our future, "The whole 20th century, because we've been speeding up to this point, is equivalent to 20 years of progress at today's rate of progress, and we'll make another 20 years of progress at today's rate of progress equal to the whole 20th century in the next 14 years, and then we'll do it again in seven years. And because of the explosive power of exponential growth, the 21st century will be equivalent to 20,000 years of progress at today's rate of progress, which is a thousand times greater than the 20th century, which was no slouch to change." (Kurzweil & Meyer, 2003). Technology triggers massive change, and massive change triggers massive fear of massive change. People, and implicitly, organizations, do not render the best decisions when these decisions are made governed by fear or under duress. With a collective resistance to technology-driven change, procrastinating against transforming with the times until it is way too late, and therefore done under duress, organizations through recent history have demonstrated instructive use cases of what we might expect more of, should we not learn from these lessons. As President John F. Kennedy famously stated "The time to repair the roof is when the sun is shining".

Change Response Behaviors

Understanding change at a broader level, and more importantly comprehending the impacts of the change, and how to navigate it is crucial for organizations - and implicitly, for the people who are the decision-makers that constitute these organizations. Simply understanding that change is happening is inadequate for people and organizations to survive and thrive. The skills to rapidly understand change and evolve ourselves and our organizations with agility is a core skill that each of us will need to have in the 21st century. Being able to make these swift pivots, the ability to fail fast and learn faster, and being able to continually learn are skills that will prove to be the difference in how successful we will be in the next twenty-five or so years. The past twenty-five years once again prove that past is prologue for the next twenty-five. Let's explore some instructive and illuminative case studies from the corporate world over the past twenty-five years. These corporate case studies examine organizations that faced rapid changes, and in response to these rapid changes, either:

a. Chose to "scream" and make a series of poor decisions under duress on the road to obscurity,

b. Or, chose to ride out the change and bury their heads in the sand,

c. Or, chose to go on "cruise control",

d. Or, successfully used the change to self-disrupt and transform themselves.

i. Some screamed

"Neither RedBox nor Netflix are even on the radar screen in terms of competition." - Blockbuster CEO Jim Keyes, 2008.

Those unfamiliar with Blockbuster, either because they are young enough to not know a world without streaming services, or were a part of the transitionary wave between Blu-ray and streaming, cannot fully appreciate the ritual that used to be a trip to the neighborhood Blockbuster. Heading to a Blockbuster store, looking for those coveted new movie titles releases, finding a copy of the movie before it was rented out, and standing in line to checkout, getting popcorn and candy, was an integral part of most people's week. In the days before corporate email followed you everywhere through your weekend, people used to head to a Blockbuster store and rent VHS tapes – and later DVDs and Blu-ray discs – to enjoy Friday or Saturday evening movie night. The main rule at Blockbuster was that you had to return new releases within two days and other rentals within a week to avoid being penalized with late fees (fees that were later relaxed in the waning days of Blockbuster to entice customers to continue using their service). The "soft rule" at Blockbuster during the VHS tape era was to rewind the tapes before returning them in the "tape return drop-box" where many people rushed to before midnight to avoid late fees, having forgotten to return their rentals in a timely manner. "Be Kind, Rewind" was the slogan Blockbuster popularized in this era.

Consider our options with streaming media era today. Hundreds of thousands of movies and TV show titles available at our fingertips 24x7x365 without having to have to drive to a store in the dead of winter on a snowy day to rent a title, or return one to avoid late fees. Between Netflix, HBO Max, Amazon Prime, Hulu, Peacock, and the several dozen other streaming services, viewers today face a different choice than they did during the Blockbuster days. The Blockbuster days allowed for a limited selection of entertainment that the entire household – or portions of the household – could watch at any given time. Streaming allows for seemingly unlimited concurrent choices, allowing each person in a household to watch a different show or a movie.

At its peak in 2004, Blockbuster consisted of 9,094 stores and employed approximately 84,300 people, 25,800 internationally, and 58,500 in the United States. Most towns across America had at least one Blockbuster store, and these stores were an anchor store for many strip malls across the United States. As of 2022, there was one Blockbuster store still operating in Bend, Oregon, a town about 160 miles south of Portland, Oregon. Known as the "Last Blockbuster", and owned by Ken Tisher, this Bend, Oregon location has become a tourist destination since Dish Network, the owner of the Blockbuster trademark, no longer grants new franchises with the Blockbuster name, cementing this store's status as the last Blockbuster. From 9,094 stores in 2004 to one special attraction store, Blockbuster is an unfortunate example of a company that did not choose to adapt to the macro changes around them, despite having had multiple opportunities to do so. The movie rental company, Redbox, exacerbated the challenges Blockbuster faced in the early 2000s. Other challenges that were eating away at Blockbuster's business were the rise of Netflix, the economic challenges posed to consumers by "Great Recession", and the rise of other streaming services. Redbox, which still exists as of 2022, established kiosks outside anchor stores such as drug stores and grocery stores, allowing consumers to rent DVDs/Blu-ray discs via kiosks without any need for membership dues (as was the case with Blockbuster), or the need to walk into a store.

The demise of Blockbuster can be attributed to a myriad of factors, however, the rise of Netflix and Netflix' decision to disrupt their own operating model by transforming themselves into a streaming media behemoth, is likely a primary driver. Most people are surprised to learn that Blockbuster actually had multiple opportunities to purchase Netflix. The first opportunity came when Netflix was still primarily a DVD-in-the-mail distributor, using the US Postal Service to mail DVDs to customers. Customers used to navigate to the Netflix website and reserve movie titles that would then be mailed to them as a rental. Blockbuster was presented with another opportunity to acquire Netflix when Netflix themselves, recognizing the shift to streaming media (with the increasing commonality of high-speed internet), "grandfathered" their mailing physical media model towards becoming a primarily streaming service. Evolving rapidly into a 100% streaming service and offering a superior recommendation engine, Netflix became the dominant streaming media service in the world. Image 16 below is a visual of how quickly Netflix' revenues dominated Blockbuster. While Blockbuster's revenues precipitously dropped from a high of $6B in 2004 to Chapter 11 Bankruptcy, Netflix soared to a revenue of $2.2B in approximately six years in 2010, to almost $30B as of 2021 – significantly larger than Blockbuster's revenues at their peak.

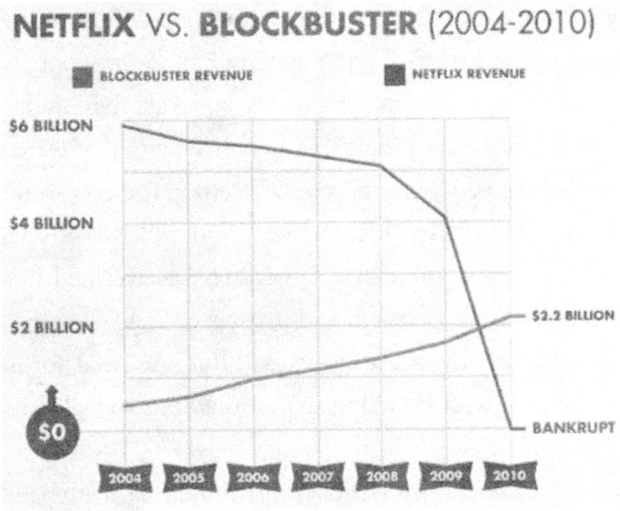

Image 16: Netflix vs Blockbuster revenues 2004 – 2010 ((Digital.net, 2011))

Netflix is a shining example of a company that self-disrupted, in anticipation of an emerging technological revolution, in time such that they could capitalize on this change. As a matter of fact, Netflix can be considered as the gold standard on how to quickly and effectively scale a global behemoth as Image 17 below depicts. With an established streaming service presence in the United States in 2007, the company quickly capitalized on the availability and rise of cloud services (ironically Netflix is hosted on Amazon's Amazon Web Service – AWS – cloud; the same company that Netflix competes with in the streaming media wars – Amazon Prime Video). Offering streaming services across the Americas and Europe in 2016, as of 2020, Netflix is available across the world (for the most part), offering a consistent and familiar experience, with regionalized and localized content wherever location you travel to. Between their superior recommendation engine, having popularized binge watching, and developing original content, Netflix continues to be the top streaming service in 2021, followed by Amazon Prime Video.

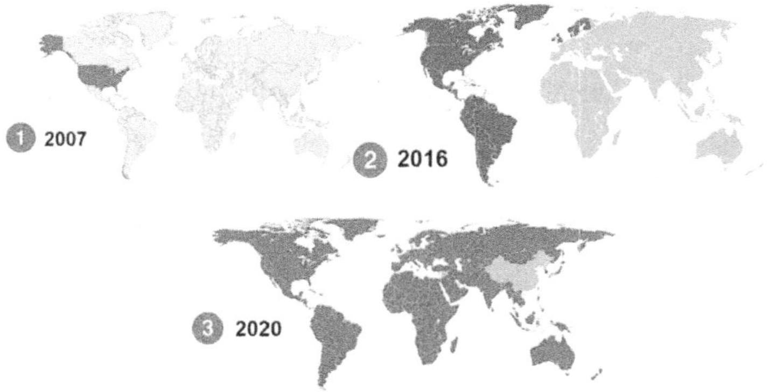

Image 17: Netflix' stunning growth in 13 years

ii. Some buried their heads in the sand

Research In Motion's flagship brand, the BlackBerry ceased service as of January 4th 2022 (NYT, NYT Technology, 2022). We discussed the iconic BlackBerry and its demise in detail in Chapter Five. Just as with the example of Nokia that we will review next, the rise of the iPhone served as a catalyst to disrupt and disintermediate the BlackBerry's dominance as the corporate mobile device of choice. BlackBerry, facing some competition from Microsoft in the corporate mobile telephony sector, made a series of egregious errors in their business model that set them on a path to obsolescence.

Three of these errors stand out the most amongst them. First and foremost, BlackBerry failed to anticipate, see, and meet the transition of the smartphone market consumer base. Starting with corporate clients, smartphones quickly found exponential popularity with consumers. This shift accelerated with the iPhone and their app-based ecosystem (the iTunes Store and the App Store). BlackBerry continued to stay focused on the corporate consumer as their customer base, and slowly but surely, corporate consumers – who were also consumers of personal electronics – moved away from BlackBerry to smartphones – devices where they could still have access to their corporate email, but could do a multitude of other things. These other things included the ability to legally purchase and download music, install games, avail of productivity apps, and even install the iBeer (Gallagher, 2022) app, which one might consider silly today, but was one of the most popular App Store downloads of its time, an app that its creators earned $20,000 from each day during its heyday. Having a mobile computer, always connected to the internet, consolidating business (email) and pleasure (entertainment and social media), was a compelling reason for business consumers to migrate from the BlackBerry to the iPhone (and later, any Google Android powered device) in droves. The second contributing factor behind BlackBerry's descent into obsolescence had to do with how smartphone manufacturers, with Apple leading the charge, created a self-contained ecosystem around them.

The Apple iOS operating system allowed for hyper-portability of user profiles, allowing users to own multiple Apple consumer electronics, such as the iPad and the iPhone, as well as the MacBook or iMac, while also creating an entire industry of "App Developers" that built software and listed it for download or purchase on the App Store, with Apple's approval. This allowed Apple (and later, Google with the Google Play Store) to create an entire self-contained ecosystem where consumers purchased and/or downloaded apps that allowed them to find entertainment, productivity, and social activities. BlackBerry had no such ecosystem, choosing instead to focus on corporate email and a browser, limited by its ability to deliver the same experience that iPhones did. It is noteworthy that in the late 2000s, mobile-friendly websites were far and few between, and the BlackBerry's browser did not deliver the same kind of responsive, engaging user experience as Apple's apps did.

The third factor contributing to Apple, and later Google, disrupting BlackBerry was that BlackBerry was too late in adopting tactile interaction. The responsive touch screen was hardly a new innovation when Apple first popularized it with the iPhone. Apple changed the course of history with touch interaction between humans and electronics. Most people over the age of 15 (as of 2022) would be astonished to know that touchscreens have been around since 1965. I say 15 because we instinctually think of the iPhone being introduced in 2007 with a touchscreen, and celebrate the iPhone as having invented touchscreens. As a matter of fact, "the first touchscreen was invented in 1965 by Eric A. Johnson who worked at the Royal Radar Establishment in Malvern, England. His first article, "Touch display—a novel input/output device for computers" describes his work and features a diagram of the design" (Displays2go, 2021). BlackBerry, having popularized their phone QWERTY keyboard, and having introduced the "medical condition", the "BlackBerry thumb" into the lexicon, resisted the marked shift towards touchscreens – a technology that Apple brought into the mainstream. As of 2022, the vast majority of smartphones – Apple, Android, and other, are touchscreen devices; devices that are "touched" an astonishing average of about 3000 times a day (Naftulin, 2016).

iii. Some went on "cruise control"

While some companies try to ignore the changes that abound around them, expecting these changes to simply pass by them, and treat them as passing fads, others end up resting on their laurels and completely miss opportunities to transform. This could be said of Blockbuster, when the firm was at its peak in 2004, or of Research In Motion (BlackBerry), but Nokia is emblematic of an organization going on "cruise control". In 2007, just before Apple introduced the iPhone, Nokia had a significant market share of the mobile phone market. Between the BlackBerry and Nokia, it is fairly clear that the Apple iPhone disrupted the entire mobile phone marketplace. How dominant was Nokia? Image 18 (Nokiamob, 2017) below illustrates how dominant Nokia was in 2007. Forbes magazine featured former Nokia CEO, Olli-Pekka Kallasvuo, on the cover of the November 12th 2007 edition. With their CEO cradling a Nokia series 6 flip-phone, the cover title read: "Nokia, ONE BILLION CUSTOMERS – CAN ANYONE CATCH THE CELL PHONE KING?" Nokia was indisputably dominating the mobile phone market in 2007, and this dominance made them reticent to acknowledge the palpable shift in the industry from cellphones to smartphones. Fifteen years after this edition of Forbes came out, the iPhone continues to be one of the most popular smartphones in the world, whereas Nokia completely withdrew from the cell phone market, after having selling their cell phone business to Microsoft.

Image 17: Forbes Magazine Cover – November 12th 2007 (Credit: (Nokiamob, 2017))

iv. Others successfully embraced the change and transformed

While history is full of illustrative case studies of organizations that failed to anticipate, understand, survive, and thrive through change, there are several examples of organizations that disrupted and transformed themselves to achieve immense benefits. Image 18 below is the cover of Business Week magazine in February of 1996. The headline "The Fall of an American Icon" heralded the death of Apple Computer. Up until 1996, Apple had been focused on the personal computing market. Apple computers - the Macintosh – had targeted corporate users as well as strove to be a household computer. Competing against Microsoft and popular (and revolutionary in its own right) Windows 95 operating system, and Bill Gates' vision to place "a PC in every office and in every home", Apple Computer was in deep trouble. A full eleven years after his firing, Apple rehired Steve Jobs as CEO in December of 1996. In what stands as a testament demonstrating that eliminating your competition is not always the only way to win, Bill Gates, on a request by Steve Jobs, shockingly invested $150M in Apple in August of 1997. Gates' investment effectively saved Apple, which at the time was on the brink of bankruptcy. Speaking about the investment at a conference forum with Bill Gates in 1997, Jobs' said "Apple didn't have to beat Microsoft. Apple had to remember who Apple was because they'd forgotten who Apple was." And with that statement, Jobs' pushed Apple back to its roots of innovation and creating demand for new products and services that consumers did not know that they wanted at the time. Jobs' invested in developing "cooler" computers, building on the cult-favorite status that the Macintosh had achieved, and introduced the iMac in 1998. Jobs and Apple's biggest moves in this era was to diversify from their sole focus on personal computing to consumer electronics. Development of the iPod and subsequently the iPhone and the iPad, centered on the iOS operating system and revolving around their ecosystem with iTunes and the App Store, Apple revolutionized multiple industries and established whole new industries around them with these moves.

BusinessWeek's cover, "The Fall of an American Icon" heralded what at the time was the imminent demise of Apple. 22 years after this cover came out, based on their capitalizing on a changing marketplace, Apple reinvented themselves and became a trillion dollar company. And they weren't done. Fueled by new products such as the Apple Watch, the AirPods, and continual updates to the iPhone and iPad, Apple, in August of 2020 became a TWO trillion dollar company. But they still weren't done. Apple's market value reached an astonishing THREE trillion dollars in February of 2022. It is remarkable for a company on a verge of bankruptcy in 1997, to reach a market value of three trillion in 2022, just twenty-five years later based on having anticipated the change around them and thriving in the midst of it.

Image 18: Cover of BusinessWeek in February of 1996

In our journey thus far, we have oriented ourselves to the fact that technological change is accelerating exponentially all around us, learned of multiple cases through history where complacency and resistance to change led to obscurity, and surmised that our own hardwired resistance to change complicates our ability to adopt and adapt. We know that the 21st century is going to need us to adopt and adapt faster than we have ever before as a species. We understand that unless we embrace a learn-unlearn-relearn mindset, we might be threatened with disruption and career displacement. The next several chapters, starting with further and clear evidence that no profession is safe from potentially significant disruption, shall outline tips and techniques for us to be able to overcome our own bias against change and ensure our "**thurvival**" in the "Fourth Industrial Revolution".

SECTION FOUR

UNDERSTAND HOW TO BE READY FOR THE CHANGE

HOW TO THURVIVE

Chapter Thirteen - Survive

Over the past 25 years, most of us have had the luxury and liberty of sitting back and observing change disrupt, and at times, decimate companies and entire industries. We've had the ability to evaluate the decisions these companies made and "Monday morning quarterback" these decisions as we read about them in college textbooks, essays, case studies, books, podcasts, YouTube videos, and through many other forums. We have been able to look back and castigate their leaders and their judgment on their lack of sound judgment amidst rampant change that threatened their companies. While it is easy to reflect on someone else's decisions, not having been there at that time, not having been faced with the same dilemmas, not dealing with the decisions they made, it is also painfully clear that the poor decision-making of these firms and their leaders were directly a result of their inability to foresee the changes that would certainly disrupt them and their unwillingness to change to survive and thrive amidst these changes.

Whether it was the employees of Blockbuster, Radio Shack, Circuit City, Toys R Us, Sears, Kodak, or any iconic American retailer, we expressed sympathy to those impacted. Whether it was loss of business for taxi cab operators, hotel staff, servers and wait staff, we understood the plight of those disenfranchised. Regardless of what sector of the economy felt the opening salvo of technology-fueled change within their industry, we were empathetic about the loss of income and disruption of livelihoods of so many millions of our fellow citizens. However, unless you were directly impacted - or your community was directly impacted (such as when Rochester, New York suffered a downturn in the local economy when Kodak stumbled) – these disruptions were always observed from a distance. Over these past twenty-five years, unless you were part of the leadership team of the organizations making these poor decisions, an employee who was impacted as a direct result, or in an industry that supported or supplied to these companies, we were able to sit back and bear witness to company after company being relegated to obscurity without our careers and livelihoods being directly impacted. While we felt the downfall of these companies in some shape or form, the demise of these titans did not jeopardize our ability to pay the mortgage and rent or put food on the table.

Most of us would continue to be comfortable in the status quo, reticent to proactively change or have to change. We believe that our livelihoods are safe. This belief might have certainly been truer forty to fifty years ago than it is in the 21st century. Fifty years ago, you could graduate high school and secure a skilled trade. This skilled trade would then likely be your vocation for the entirety of your working life. Some of those in these skilled trades segued into leadership and management jobs within those professions. Fifty years ago, you could graduate college, secure an entry-level job across any field, and that would be your career. You would likely retire with the very same company that you started with, more than likely with a pension benefit. Your career would witness the ups and downs of your organization, and you would have a front row seat to the changing nature of the business and technological advancements that helped augment your job. You might embrace some of these advancements, be unwillingly mandated into others, but by-and-large, your job - and implicitly - your livelihood - was secure. It would be unusual for people to have to switch careers and fields. Heck, it was unusual for people to even switch companies back then.

Contrast that to today when a single employee belonging to Generation Z is expected to change jobs up to 10 times between the ages of 18 and 34 based on a 2022 survey of US students (Zurich, 2022). It is expected that the same Generation Z employee will likely have five careers and an average of 18 jobs through these five careers. Whereas some mobility allows individuals to accrue disparate experiences and therefore continue to grow and develop in their careers, at what point does the need to gain breadth of experiences dilute and detract from the ability to learn a job in depth? Now add onto it the fact that most of the jobs that those in Generation Z might hop across will continue to stretch and contort as technological pressures force the very nature of these jobs to change. How does one develop subject matter expertise in any one domain or subject area under such fluid and malleable career conditions? What if you belong to the part of Generation X that would be eligible for retirement in the twenty-five year timeframe of the future we have been examining throughout this book? You have likely had several jobs across multiple companies through your career. You have built up a rich career, chock full of a lifetime of experiences. What happens if your job is eliminated due to technological change when you are at age 60, five years shy of retirement? With all your retirement plans based on you earning until age 65, this could be a severely financially damaging event. What if that happens at age 55 when you still have a solid decade of service more to go? How disruptive, damaging, and potentially financially devastating can something like this be?

Whether you are 25 or 52, it is vital for each of us to prepare ourselves for the pronounced changes in existing occupations, the elimination of some existing occupations, and the opportunities that will be abundant in this new hyper-digital economy of the 21st century. Before we explore how exactly we can go about doing that, it is worthwhile to understand the shifting occupational landscape.

Disappearing Professions - 2030

According to a 2021 online article that was published in the midst of the COVID-19 global pandemic, "In 1950, the job of elevator operator was among the 270 careers listed on the United States Census. That job title is now extinct, representing the only known instance of an entire occupation being obliterated by automation in the 50 years that followed. The next half-century may be less forgiving. Sophisticated software, robotics, automation, artificial intelligence (AI), and changing trends threaten the livelihoods of everyone from taxi drivers and restaurant servers to computer programmers and librarians. Many economists predict that automation, not outsourcing, will lead to the loss of more than 1.5 million jobs in America's manufacturing sector. These technical innovations will soon render many longstanding skills and trades obsolete—and the occupational Grim Reaper will discriminate according to class.

Many of the jobs most likely to disappear are among the last well-paying jobs one can get with only a high school diploma. Low-paying, unskilled jobs with low educational entry barriers are most susceptible to automation. These are the jobs that robots will do. Manufacturing will require greater technical skills to operate and program computers. Those who lose their jobs will largely be shut out of the high-paying, highly skilled jobs that remain, many of which will go to specialists tasked with tending to and improving upon the very machines and programs that replaced the human workers." (Lisa, 2021).

There will always be individuals who believe that they are impervious to any technological change happening around them. That their industry is safe. That specifically, their own job is safe and insulated from technological disruption and change. However, there isn't a single job across any sector of any industry that can claim to be 100% safe and shielded from change over the next twenty-five years with any level of certainty. Those who claim otherwise are deluding themselves to becoming inert to change, or worse, influencing others that they shouldn't have to change. There will be no job - **none** - that will be not impacted as a result of the "Fourth Industrial Revolution" washing over industries, sectors, companies, and communities like a continually cresting and unrelenting technology tidal wave.

As of this book's publication in 2023, people can author computer code using AI, write entire books or engaging essays using ChatGPT (a chatbot launched by OpenAI in November of 2022), conduct medical exams remotely using a combination of data collected by a smart device and via a Zoom call with a physician (telemedicine), vacuum your entire home using a robotic vacuum clearer like the Roomba, and mow your lawn using a robotic lawn mower. From landscaping to medicine, no profession is safe from disruption. None of these disruptions are a bad thing – they are absolutely required for our continued progress – we just need to do our parts to stay a step ahead, or at the least, be able to keep up.

Adopted from articles appearing on the AARP website (Terrell, 2022), which was based on data collected from the US Bureau of Labor Statistics (BLS), and CareerAddict, here are a few professions that are projected to lose the most number of professions by 2030, at the end of this decade:

1. Parking enforcement workers: With total number of jobs in at 7900 positions, projected to be at 5100 positions by 2030 for a 35% loss of jobs, the article states, "Many things have changed since the first parking meter was installed on July 16, 1935, in Oklahoma City. While parking tickets won't be going away anytime soon, there will be fewer workers handing them out. Digital parking apps have made it easier to spot which meters are running low, creating less need for parking enforcement officers on patrol." (Terrell, 2022).

2. Telephone switchboard operators: With total number of jobs in at 4800 positions, projected to be at 3600 positions by 2030 for a 25.4% loss of jobs, the article states, "Once upon a time, telephone operators were essential workers. Instead of calling another phone directly, you needed an operator to connect you. For many years, it was a booming profession — especially for women — with as many as 235,000 operators in 1930." (Terrell, 2022).

3. Watch and clock repairers: With total number of jobs in at 2800 positions, projected to be at 2100 positions by 2030 for a 24.9% loss of jobs, the article states, "With most people using their cellphones or other digital devices to tell time, the demand for watch and clock repairers is ticking away. Watches do continue to be popular as luxury items and heirlooms, though, so these jobs may not disappear entirely." (Terrell, 2022).

4. Door-to-door sales: With total number of jobs in at 54,000 positions, projected to be at 41,000 positions by 2030 for a 24.1% loss of jobs, the article states, "Demand for workers in this profession is falling fast due to the rise in direct marketing through email and the internet. But talented salespeople certainly can still find work in other related professions." (Terrell, 2022).

5. Bank Tellers: With total number of jobs in at 432,500 positions, projected to be at 359,400 positions by 2030 for a 16.9% loss of jobs, the article states, "Since the moment the first ATM opened on Sept. 2, 1969, in Rockville Centre, New York, the career prospects of bank tellers have been gloomy. With more people using banking apps and cash apps such as Venmo, PayPal and Apple Pay now - that means even fewer people are going into banks to make transactions." (Terrell, 2022).

6. Assemblers and Fabricators: "When it comes to occupations that include manual labor, automation seems inevitable. Assemblers are mainly responsible for making commodities such as toys, vehicles and aircrafts, among other products. However, with machines and robotics taking over assembly processes and repetitive tasks, assemblers are becoming an endangered species in most industries. According to the US Bureau of Labor Statistics, the profession will face an 11% decline by 2028, resulting in the loss of some 203,300 jobs." (Phillpott, 2021).

7. Travel Agents: "With the abundance of easy-to-use comparison websites, anybody can arrange their own holiday. All you need is your bank card and a few spare hours to research your destination, with the likes of Skyscanner, Trivago and Opodo tailoring flight and hotel searches to your exact price and date range. Many travel operators have realized this and are closing down branches to focus on their online offers." (Phillpott, 2021).

8. Cashiers: "The reality of a cashless society, with advances in contactless payments, Apple Pay and even cryptocurrencies such as Bitcoin are becoming prominent within mainstream society. While not everyone is on board, with some preferring to still use cash to better track their spending, one thing is for sure: the requirement for people to handle the payments is no more. With self-service stations already a common site in supermarket chains and even fast food restaurants like McDonald's, the demise of the cashier seems inevitable." (Phillpott, 2021).

...And 50 Years Out

Thus far, we've taken a modest look forward to surveil the job market of the future for what types of jobs will face declining prospects through the end of this decade. We've established that there are clearly a diminishing number of professions that might be absolutely insulated from significant technology-driven disruption. This birds-eye view perspective is important for us to ground ourselves. We need to clearly understand the stakes. We need to fully realize that the comfort zone we tend to occupy is a barren landscape where no career growth can happen. If the projected job losses at end of this decade were not indicative enough, the next fifty years will bring immense change across sectors and industries. The next fifty years shall present incredible opportunities and new career prospects, all the while disrupting jobs and professions as we know it. Consider that a hundred years ago, there were no jobs at scale that dealt with anything electric (outside of wiring America with the electric grid itself). None. Zero. With less than half of US households powered by electricity, there was no industry around consumer electronics. No computers, no TVs, no affordable refrigerators, no microwaves, no electric incandescent lights, and so on. These types of jobs simply did not exist. Heck, these industries and sectors simply did not exist.

Think of the technology industry as we know it today. The Fortune 100 list is dominated by technology companies. These companies employ hundreds of thousands of people across the globe. None of these jobs existed twenty-five to fifty years ago. There was no profession called "software developer", or "video game developer" fifty years ago. Who would have thought that a job called a "YouTube Influencer" would actually be a thing, when YouTube itself did not exist twenty-five years ago? The "Fourth Industrial Revolution" will usher in an era where there will be new sectors and new industries created that do not exist today – an era that will witness the creation of hundreds of thousands of new jobs and professions that we cannot imagine today. At the same time, a lot of the jobs that provide reliable employment today will likely look significantly different in fifty years, or cease to exist in their entirety. Some of the jobs that will likely be challenged over the next twenty-five years and face extinction in fifty are:

1. Passenger Transport Drivers: The continued growth of autonomous driving vehicles and autonomous aerial vehicles like the ones we explored in the first chapter, will pose a significant challenge to the taxi cab industry, and specifically, the taxi cab driver profession. Those who drive for Uber, Lyft, and other car ride sharing services will also be severely challenged to exist. The same will happen to bus transportation and bus drivers.

2. Commercial Drivers: Following a road similar to passenger transport drivers, commercial drivers, such as those who are an integral part of the supply chain, transporting goods and products to point of sale, as well as long haul truck drivers who transport materials all across the United States, will face disintermediation due to self-driving trucks.

3. Pilots: Similar to drivers of passengers and products, it is highly likely that air travel will see a transition from human pilots to pilotless autonomous airplanes in the next fifty years. This transition will likely be commensurate to the shift to eVTOL craft as air taxis. Even though they might be

introduced, commercial aircraft transporting hundreds of passengers might face slower adoption than air taxis would.

4. Traffic and parking enforcement: It is highly likely that traffic and parking enforcement would transition from human enforcement officers to a combination of drones and robots.

5. Fast food workers: Fast food workers effectively operate as assembly line workers, and assembly lines are prime targets for robots and automation. It is highly likely that fast food workers will unlikely be a job available to high school students in fifty years.

6. Retail Cashiers: The number of staffed cashier lines continue to be reduced in favor of self-checkout counters across most retailers. It is highly likely that cashiers will be entirely eliminated in the next fifty years, posing yet another challenge to entry-level employment for high school students.

7. Translators: Imagine all the human translators employed today across sectors – both in the public and private sectors. With the increasing sophistication of technology-based translators and translations services, it is foreseeable that this profession will cease to exist in the next fifty years.

8. Stockers and Warehouse Workers: Enter any Amazon distribution center today, and you will witness significant automation and robotics at work. It is highly likely that stockers and warehouse workers will be entirely replaced by robots in the next fifty years.

9. Customer Service Operators: Most organizations' customer service centers today offer an omni-channel way for customers to connect the organization for service – online and via the phone. Most organizations also ensure that it isn't easy to connect with a real live human on the customer service desk. Live humans are the most expensive aspect of operating

customer service centers. Customer service organizations also increasingly employ bots – from sophisticated chatbots that can interact with customers online and on mobile apps, to conversational AI that can have an engaging dialog with customers. Conversational AI will continue to mature and will become indistinguishable from live humans in the next decade, posing a threat to the viability of this profession in the next fifty years.

10. Factory workers: Most repeatable tasks that rely on humans today will be automated by technology, leveraging a combination of AI and robots in the next fifty years. Factory workers and assembly line operators, those performing any tasks that can be automated, will most certainly be fully automated agnostic of industry or sector.

AI, AI, AI

AI will more than likely be leading the charge in terms of all the technology-powered change across sectors and industries. Implicitly, of all the industries and professions that will emerge (and be disrupted) in the next few decades, AI-related industries and professions would likely see strong growth. The 1920s were the age of electrification expanding across America and the world. The 2020s will be remembered as the age of AI expanding across America and the world. Every industry in every sector is well-positioned to capitalize on the benefits of this expansion of AI – operational efficiencies, automation, business agility, reduced costs, etc. AI will also immediately disrupt manual and repeatable tasks as discussed in the preceding chapters. We haven't even scratched the surface of the tip of the tip of the iceberg when it comes to the utility of ChatGPT and Generative AI – a phenomenon that expanded like wildfire in early 2023.

We have no real playbook for AI. The field is so new, so untested, so fluid, and changing so rapidly, that developing such a playbook today ensures that it will be outdated tomorrow. The explosive growth of

Generative AI in early 2023 and the resulting "AI Arms Race" led to dozens of industry professionals - including technology billionaire Elon Musk, Apple co-founder Steve Wozniak, and several AI researchers - calling for a six month pause on all AI-based development that was any more advanced than the version of ChatGPT that took the world by storm in early 2023. These professionals argued in an open letter that this six month pause would allow industries (and countries) to design and implement robust safety measures around the safe and effective use of AI. The open letter signed by luminaries was issued by the not-for-profit organization "The Future of Life Institute", which states that its mission is to "steer transformative technologies away from extreme, large-scale risks and towards benefiting life" (Future of Life Institute, Retrieved 2023). The letter, warning of the risk that uncontrolled AI could flood information channels with misinformation and automate countless jobs almost overnight, states that "AI systems with human-competitive intelligence can pose profound risks to society and humanity", and that "recent months have seen AI labs locked in an out-of-control race to develop and deploy ever more powerful digital minds that no-one - not even their creators - can understand, predict, or reliably control".

Adding fuel to the fire to this open letter came the resignation of Dr. Geoffrey Hinton from Google. Dr. Hinton, 75 years old at the time of his resignation from Google in late April of 2023, is widely seen as the godfather of artificial intelligence (AI). Stating that he has regretted his work, Dr. Hinton was quoted in a BBC article that the danger of AI chatbots were "quite scary" (Kleinman & Vallance, 2023).

Dr. Hinton, in this BBC article, continued, "Right now, they're not more intelligent than us, as far as I can tell. But I think they soon may be." (Kleinman & Vallance, 2023). An excerpt from this article goes on to state "Dr Hinton's pioneering research on neural networks and deep learning has paved the way for current AI systems like ChatGPT. In artificial intelligence, neural networks are systems that are similar to the human brain in the way they learn and process information. They enable AIs to learn from experience, as a person would. This is called deep learning. The British-Canadian cognitive psychologist and computer scientist told the BBC that chatbots could soon overtake the level of information that a human brain holds. "Right now, what we're seeing is things like GPT-4 eclipses a person in the amount of general knowledge it has and it eclipses them by a long way. In terms of reasoning, it's not as good, but it does already do simple reasoning," he said. "And given the rate of progress, we expect things to get better quite fast. So we need to worry about that." In the New York Times article, Dr Hinton referred to "bad actors" who would try to use AI for "bad things". When asked by the BBC to elaborate on this, he replied: "This is just a kind of worst-case scenario, kind of a nightmare scenario." You can imagine, for example, some bad actor like decided to give robots the ability to create their own sub-goals." The scientist warned that this eventually might "create sub-goals like 'I need to get more power'". He added: "I've come to the conclusion that the kind of intelligence we're developing is very different from the intelligence we have." We're biological systems and these are digital systems. And the big difference is that with digital systems, you have many copies of the same set of weights, the same model of the world. And all these copies can learn separately but share their knowledge instantly. So it's as if you had 10,000 people and whenever one person learnt something, everybody automatically knew it. And that's how these chatbots can know so much more than any one person." Make no mistake, we are on a speeding train right now, and the concern is that one day it will start building its own tracks." (Kleinman & Vallance, 2023).

An article published in MIT Technology Review during the same week added more nuance to Dr. Hinton's ringing the alarm bells on the very technology he helped create. According to this article, "Hinton says that the new generation of large language models - especially GPT-4, which OpenAI released in March - has made him realize that machines are on track to be a lot smarter than he thought they'd be. And he's scared about how that might play out. "These things are totally different from us," he says. "Sometimes I think it's as if aliens had landed and people haven't realized because they speak very good English."

For 40 years, Hinton has seen artificial neural networks as a poor attempt to mimic biological ones. Now he thinks that's changed: in trying to mimic what biological brains do, he thinks, we've come up with something better. "It's scary when you see that," he says. "It's a sudden flip." Hinton's fears will strike many as the stuff of science fiction. But here's his case.

As their name suggests, large language models are made from massive neural networks with vast numbers of connections. But they are tiny compared with the brain. "Our brains have 100 trillion connections," says Hinton. "Large language models have up to half a trillion, a trillion at most. Yet GPT-4 knows hundreds of times more than any one person does. So maybe it's actually got a much better learning algorithm than us."

"If you or I learn something and want to transfer that knowledge to someone else, we can't just send them a copy," he says. "But I can have 10,000 neural networks, each having their own experiences, and any of them can share what they learn instantly. That's a huge difference. It's as if there were 10,000 of us, and as soon as one person learns something, all of us know it."

What does all this add up to? Hinton now thinks there are two types of intelligence in the world: animal brains and neural networks. "It's a completely different form of intelligence," he says. "A new and better form of intelligence." That's a huge claim. But AI is a polarized field: it would be easy to find people who would laugh in his face—and others who would nod in agreement. People are also divided on whether the consequences of this new form of intelligence, if it exists, would be beneficial or apocalyptic. "Whether you think superintelligence is going to be good or bad depends very much on whether you're an optimist or a pessimist," he says. "If you ask people to estimate the risks of bad things happening, like what's the chance of someone in your family getting really sick or being hit by a car, an optimist might say 5% and a pessimist might say it's guaranteed to happen. But the mildly depressed person will say the odds are maybe around 40%, and they're usually right."

Which is Hinton? "I'm mildly depressed," he says. "Which is why I'm scared."" (Heaven, 2023).

As one might expect, Dr. Hinton ringing the "AI alarm bells" - soon after dozens of prominent technology professionals having also done so – set off a worldwide media frenzy on the safety and efficacy of AI. While there were a handful of skeptics who questioned if AI had truly become that advanced that quickly, there were others who believed that we were already too late – that AI had already expanded rapidly in areas that we knew, but more concerning was their speculation that AI had already expanded in areas that we were unaware of. While I am a firm believer in the benefits from AI – to companies and to communities – there are some real concerns and quandaries that many sectors are wrestling with in 2023 with respect to this technology. The ethics of AI and around AI are a prominent concern across many industries. The safe and effective use of AI by virtue of a framework known as Explainable AI (XAI) seeks to solve some of the inherent challenges with ungoverned AI. Some of these concerns include the potential lack of transparency of how an AI algorithm arrived at the decisions that it did arrive at, and the sheer near impossibility that a human being will be able to comprehend and retrace the decision-making pathways and processes that a machine that is capable of millions of computations a second can make.

If humans are unable to understand how an AI arrived at a conclusion or recommendation that it did, it becomes very difficult to trust in the decision. Regulated industries such as healthcare and insurance have to be particularly watchful to ensure they are complying with emerging regulations intended to protect the consumer. The use of Generative AI and the implications of cheating in academia is one such concern with ethics and AI, but there are very real concerns that AI might make decisions that are inadvertently biased or accidentally discriminatory in nature. These kinds of challenges can occur due to inherent quality of data being used, or by the underlying AI and machine learning (ML) models themselves.

Additionally, if an AI model has been provided instructions to disregard specific data elements or attributes, the model will always disregard these traits. However if the model does not have just as specific instructions to ignore certain other related traits, the model might interpret these other traits that result in unintentional bias. AI models can develop unintentional bias either by the accidental reinforcement of historical discriminatory patterns, or via discovering new discriminatory patterns via unexpected correlations.

Of these two, it is easier to identify and rectify historical discriminatory patterns, whereas it is more challenging to identify unexpected correlation since AI models consume vast amounts of data to establish connections, and could and accidentally correlate unexpected data attributes. Consider a hypothetical scenario where students across the United States are applying for a prestigious Ivy League university. Imagine that this determination was being driven by an AI model, and that this AI model leveraged geographical location as a key input. In such a circumstance, it is entirely foreseeable that the zip code/geographic location of the student applicant is highly correlated with ethnicity. In other words, zip codes can reflect the majorities of particular races that reside within that zip code. Therefore, if the student's location is highly correlated with ethnicity, it is likely that the AI model would inadvertently favor certain ethnicities over others.

The greater the volume of data and the more complex a model, the greater is the opportunity for correlation and pattern inference. It is impossible for humans to be able to analyze and discern correlations between traits that an AI model learns in contrast to what it was taught to learn. AI / ML models "self-learn" by crunching through vast amounts of data and by looking for correlations within this data. Sometimes, the systems can inadvertently correlate disparate elements of data to arrive at a prediction or recommendation. Although this correlation is unintended, it can have far reaching impacts in terms of its ultimate output. Consider these two concrete examples:

i. Amazon.com and Recruiting:

Starting in 2014, one of Amazon's machine learning teams had been developing a program to vet the resumes of job applicants. The goal was to vet the resumes in an automated manner, reduce reliance on human recruiters, and provide top talent recommendations to these human recruiters, allowing them to recover some of their capacity – while delegating some of the rote, manual work to the algorithms. Much like shoppers on the e-commerce online retailer's platform are allowed to rate products on a one through five scale, this recruitment screening AI tool scored job candidates on a five-scale.

According to a Reuter's article describing the issue, "Automation has been key to Amazon's e-commerce dominance, be it inside warehouses or driving pricing decisions." (Dastin, 2018). The article goes onto quote an Amazon employee who stated that the company "wanted it to be an engine where I'm going to give you 100 resumes, it will spit out the top five, and we'll hire those." (Dastin, 2018). After about a year of the AI being operational, Amazon realized that the system had not been rating candidates for software developer and IT jobs in a gender-neutral manner.

The AI/ML models that had been trained to evaluate candidates by pattern recognition and observations in resumes that were submitted to

Amazon over a decade. Reflecting that these professions have historically been male dominated, the model self-learned that male candidates were "preferable". The model penalized applicant resumes that included words such as "women's" as the word appeared in terms such as "women's chess club champion", downgraded graduates of two all-women colleges, etc. While Amazon "did not dispute that recruiters looked at the recommendations generated by the recruiting engine" (Dastin, 2018), the company does claim that their recruiters did not leverage the algorithm to evaluate candidates or make hiring decisions. The Amazon story has been a lesson to other organizations such as Goldman Sachs, and Hilton Worldwide Holdings, that had been planning to leverage similar utilities for their own recruitment practices.

ii. Microsoft's Chatbot:

The second example is of technology titan, Microsoft, and their experiment with Microsoft Tay. Microsoft released Tay (an acronym for "Thinking About You") on March 23rd 2016. Tay was an AI chatbot that was designed to self-learn from the internet, emulate an American teenage girl, and automatically post messages to social media site Twitter in the form of "tweets" under the Twitter handle "TayTweets". Tay was devoid of human oversight given the 'self-learning and tweet post' nature of its purpose. In less than 16 hours, Tay self-learned from the internet to become extremely racist, xenophobic, and sexist – these learning manifesting themselves in the AI's tweets. Microsoft had to take Tay offline and publicly apologize.

Amazon and Microsoft are both highly sophisticated technology titans. If these missteps could befall these digitally-native organizations, consider the challenge posed by "guardrails-free" AI use across sectors with anachronistic technologies.

"**Thurviving**" in the age of AI is a microcosm of "**thurvival**" skills required for technology as a whole, as the next several decades will witness an unprecedented convergence of technological advancements, with AI leading the charge. The opportunities for humans to thrive in this era are vast, but they come with the responsibility to adapt, learn, and evolve. By embracing lifelong learning, fostering creativity, navigating ethical implications, collaborating with AI, and cultivating emotional intelligence, individuals can position themselves to not only survive but also thrive in the face of AI-driven changes. The future is brimming with potential, and as authors of our destiny, we have the power to shape a harmonious relationship between humans and intelligent machines.

The rapid evolution of technology and AI will lead to the emergence of new industries and job opportunities. While some traditional roles may be automated, this technological shift will create a demand for human expertise in areas such as AI development, data analysis, cybersecurity, and human-AI interaction. Additionally, as AI takes over mundane tasks, there will be increased opportunities to pursue creative endeavors, entrepreneurship, and personalized skill development. In the face of AI-driven changes, the key to survival lies in our ability to adapt and embrace lifelong learning.

As technology continues to evolve, traditional educational models need to evolve to keep pace with the evolving demands of the job market. We will need to continuously update our knowledge and acquire new skills to remain relevant. Adapting to new technologies, such as AI systems, robotics, and automation, will require us to develop a strong foundation in STEM (Science, Technology, Engineering, and Mathematics) fields while also nurturing creativity and critical thinking abilities. Corporations, governments, educational institutions, and we, as individuals, must invest in reskilling and upskilling programs to equip the workforce of tomorrow with the necessary competencies required in a technology-driven society. Embracing a growth mindset and cultivating curiosity will enable us to embrace change and seize new opportunities.

As AI becomes increasingly prevalent, it is imperative to prioritize ethical considerations and ensure the development of human-centric AI systems. We – as humans - must maintain control and transparency over AI algorithms, guarding against biases, and discriminatory practices. Safeguarding privacy, data protection, and cultivating trust in AI technologies are essential for sustainable progress. Collaborative efforts between governments, industry leaders, and AI researchers should focus on establishing ethical frameworks and guidelines to ensure the responsible and beneficial deployment of AI. As AI becomes increasingly integrated into various aspects of society, addressing the ethical and social implications becomes paramount. We must actively engage in discussions around AI governance, privacy, bias, and fairness. The responsible development and deployment of AI technologies should be a collective effort involving policymakers, researchers, and the public. As authors of the AI-driven future, we must advocate for transparency, accountability, and ethical decision-making to ensure that technological progress aligns with our societal values.

Rather than fearing the displacement of human labor, we should view AI as a powerful tool that can augment our capabilities. Collaboration between humans and AI systems has the potential to amplify productivity, streamline processes, and enhance problem-solving. Learning to work alongside AI, understanding its strengths and limitations, and effectively integrating it into workflows will be key to thriving in the AI-driven era. By leveraging AI to handle repetitive tasks, we can focus on higher-level cognitive functions, relationship-building, and strategic thinking.

While AI can analyze vast amounts of data and perform complex calculations, it lacks the ability to comprehend and exhibit human emotions. Emotional intelligence (EQ) and empathy are integral to building strong interpersonal relationships, understanding diverse perspectives, and effectively collaborating. In an AI-driven world, individuals who can harness their emotional intelligence will thrive. Cultivating empathy, active listening, and emotional resilience will enable us to forge connections and lead with compassion in an increasingly digital landscape.

AI, with its immense computational capabilities and ability to process vast amounts of data, presents an incredible opportunity for human-AI collaboration. By leveraging AI's analytical prowess, we can harness its capabilities to enhance productivity, creativity, and problem-solving across numerous domains. This collaboration will enable us to tackle complex challenges more efficiently and uncover innovative solutions that were once beyond our reach. While technology and technology-based automation can streamline processes and enhance efficiency, striking a balance between automation and human involvement is crucial. Recognizing the unique strengths of human creativity, emotional intelligence, and critical thinking is essential in leveraging technology and AI as a tool rather than a replacement.

By integrating AI technologies thoughtfully, we can augment our human capabilities, freeing us to focus on complex problem-solving, innovation, and compassionate interactions. While machines excel at repetitive and rule-based tasks, they struggle with creativity and innovation. We - as humans - possess the innate ability to think outside the box, connect disparate ideas, and develop novel solutions. Cultivating these uniquely human qualities will become even more valuable in the coming decades. We should strive to hone our creative skills, embrace a culture of experimentation, and embrace failure as an opportunity for growth. By collaborating with technology, we can leverage our creativity to develop groundbreaking ideas and innovations that drive progress.

The next several decades of technological change, driven by AI, hold tremendous potential for human advancement. Embracing this future requires a proactive mindset, adaptability, and a commitment to lifelong learning. By leveraging the power of human-AI collaboration, seizing new opportunities in emerging industries, prioritizing ethical considerations, and nurturing our unique human qualities, we can not only survive but thrive in the AI-driven era.

The rise of AI and technological change undoubtedly presents career risks, but individuals who proactively adapt can find success and fulfillment in the face of these challenges. By cultivating a growth mindset, embracing lifelong learning, fostering adaptability, leveraging transferrable skills, staying attuned to emerging opportunities, and fostering a supportive network, individuals can navigate the shifting career landscape and carve out a resilient and prosperous professional journey. The future brims with opportunities for humans to thrive in the era of AI and technological change. By embracing AI as a collaborative partner, cultivating uniquely human skills, fostering creativity and innovation, prioritizing lifelong learning, and championing ethical practices, individuals can unlock their full potential. The key lies in embracing the symbiotic relationship between humans and AI, harnessing the power of technology while preserving and enhancing our distinctively human qualities. By doing so, individuals can shape a future that combines the best of both worlds, leading to unprecedented progress, fulfillment, and shared prosperity.

Being Proactive Today, Building A Promising Tomorrow

We have by now clearly established that the rapid pace of technological progress and automation will likely result in the obsolescence of many jobs that exist today over the next 25 to 50 years. Jobs that require routine, repetitive tasks are particularly vulnerable to being automated, and will likely be replaced by machines or software that can perform the same work more efficiently and accurately. We've explored several such vulnerable professions that may become obsolete. Some of the most vulnerable industries include manual labor, routine-based jobs, and certain administrative and office support roles.

With the continued development of automation and robotics, jobs that involve repetitive manual tasks such as assembly line work, packaging and labeling, and data entry are likely to become obsolete. Additionally, with the advent of artificial intelligence and machine learning, jobs that involve routine-based tasks such as bookkeeping, data analysis, and customer service are also at risk. As you can see, the professions in imminent risk of extinction include white collar and blue collar jobs alike. White collar, offices jobs such that of a data entry clerk will be extinct as advanced AI algorithms and machine learning techniques will automate data processing and analysis. Blue collar jobs in the manufacturing sector that involve repetitive manual labor that would likely be replaced by robots and other automated systems.

Thurvival for the jobs in imminent risk of being made obsolete raises the stakes for those serving in these positions today. We will learn the tools for how to **thurvive** over the next few chapters, these techniques being relevant whether your profession is in imminent danger, or whether that danger is a few years away. In order to prepare for this change, individuals in jobs that are at risk of being made obsolete need to invest in their education and skill development. This can be done through formal education, such as taking courses and obtaining degrees in fields that are less likely to be automated, or through on-the-job training and experience. Individuals must be proactive and seek out new skills and knowledge that will be in high demand in the future. This may involve upskilling or reskilling in areas such as data analysis, coding, and digital marketing. It's also important for individuals to stay up-to-date with the latest technological advancements and to be proactive about acquiring new skills and knowledge that will make them more valuable to employers in the future. It is also important for individuals to stay informed about the latest technological advancements and understand the potential implications for their current or future careers.

This can involve attending conferences and workshops, taking online courses, and networking with experts in the field. Another way for individuals to prepare for the impact of technology on their jobs is to develop transferable skills, such as critical thinking, problem solving, and creativity, that are less likely to be automated. It will be even more important in developing expertise in areas such as ethics, creativity, and emotional intelligence, which are unlikely to be easily and imminently replaced by technology. These skills are in high demand in a rapidly changing job market, and will make individuals more adaptable and competitive in the future.

The next few chapters shall explore these **thurvival** techniques a bit more in-depth. As much as we need to prepare ourselves with **thurvival** techniques for our individual selves, it will also be important for individuals and society as a whole to consider the broader implications of technological unemployment, and to consider how best to support those who are displaced by automation. Governments and organizations, through us as individuals, must also play a role in helping people prepare for the future of work.

This can include offering job training and education programs, unemployment insurance, government support for retraining and skill development, creating incentives for businesses to invest in upskilling their workforce, and developing policies that support workers who are displaced by technological change. While the automation of many jobs will likely result in significant disruptions to the labor market, we can prepare for this change by investing in their education and skill development, acquiring transferable skills, and supporting initiatives that help to mitigate the negative impact of technological unemployment. By being proactive and seeking out new skills and knowledge, we can position ourselves for success in the rapidly changing world of work.

Our **thurvival** depends on it.

Chapter Fourteen – Thrive

The COVID-19 pandemic significantly accelerated digital transformations and digitization activities that were underway across several sectors and industries. These digital transformations were especially pronounced in industries that had historically been slow to digitize. The simple rationale for this is that companies within these industries had to play catch up in a compressed amount of time to achieve a level playing field with some of the digital upstarts and disruptors within their own industries. A good example of that is within insurance and financial services in general. Retail banking was forced to adopt digitization driven by consumer demand and the number of interactions that each of us have with our financial institutions on a daily basis. Fueled by Millennials and Generation Z who preferred transacting fully online versus walking into their local bank branch, banks were leaders in digitization within the Financial Services industry. This was followed by automobile insurance, home insurance, and then finally other sectors of the Insurance and Financial Services industry such as life insurance and retirement planning. The same story played itself out repeatedly across several other industries, where, starting with one sector within an industry, adjacent sectors within that same industry followed suit in modernization.

These "digital transformations" were vastly disruptive to established, mature industries such as insurance, with some companies within those industries dating back two hundred or more years. With a considerable amount of legacy technical debt these organizations carried into the 2020s, modernization by and large, was a disruptive activity. Nipping away at the heels of these established companies were smaller, nimble, agile startups that sought to change the paradigm of how business was conducted. With no technical debt and a blank canvas when it came to implementation of processes, these startups seemed to pose a competitive threat to the established titans. Over the past five years or so, the ecosystem has brought these entities together, wherein across a myriad of industries you will find technology-native, so-called "digitally native" startups cohabiting and sharing their technological expertise with established firms. The firms can invest in these companies and harvest their technological intellectual capital, without having to disrupt their entire operations and business models. Most industries and most firms in sectors across industries have invested untold hundreds of millions of dollars in financial capital to digitize and modernize, these activities starting in earnest in the early 2010s. The pandemic, where face-to-face interactions were limited to non-existent, and societal embrace of everything online – shopping, working, schooling, etc. – thrust these digitization investments into the forefront, accelerating the already accelerated pace of industrial digital change.

Technology's effect on employment and jobs of the future vary vastly by occupation, industry, and geography. In what is a no-brainer, jobs within the technology industry will continue to grow, specifically within cybersecurity, data science, and AI. Technology jobs are used to continual change and professionals within technology expect to, and desire to, continually learn new skills. In that sense, technology sector jobs and the people who serve within it, can be considered as anomalies. Those who "came of age" in the technology industry over the past twenty-five years have already witnessed an order of magnitude more change – in how they develop software, how they deploy code, how they work within a team – than they have in the twenty-five years preceding. Unsurprisingly, this tracks to the exponential growth, sophistication, complexity, and maturity of technology in general. According to McKinsey & Company, "Overall spending on technology could increase by more than 50 percent between 2015 and 2030. About half would be on information-technology services. The number of people employed in these occupations is small compared to those in healthcare or construction, but they are high-wage occupations. By 2030, we estimate that this trend could create 20 million to 50 million jobs globally." (McKinsey Global Institute, Jobs lost, jobs gained: What the future of work will mean for jobs, skills, and wages, 2017). Keep in mind that these projections from McKinsey were from a pre-pandemic perspective.

Jobs with repeatable and rote tasks, with predictable variables and predictable outcomes, are those that provide the greatest opportunity for automation. Fast food lines, assembly lines, machinery that can be programmed, gathering, ingesting and processing data – jobs that range from a McDonald's drive through operator and line cook to a paralegal and basic bookkeeper – can all be done significantly faster and significantly cheaper by technology. Of particular note across the projections of job losses within any given sector is the fact that even though declines for a particular job might not seem particularly alarming, it is a near certainty that even those jobs within those sectors that will be created will look significantly different than they do today. Workers in those jobs will be expected to learn new skills as their job functions continue to shift due to automation. Some of these workers might find themselves being retrained into different types of roles within the same industry or within the same company.

These new roles might be a hybridized combination of some of their existing responsibilities augmented with job functions of a different role that itself has been partially automated. As we shall examine later in this book, technology – AI, automation, machine learning, etc. – will have a much lesser effect on functions that are innately human today. These include the arts, and functions predicated on inter-personal relationships and social interactions such as leading teams and managing people. It is unlikely that technology will progress to an extent where it can replicate the range of human emotions – thinking AND feeling – by the end of this decade. Occupations that are multivariate in nature, dealing with a menagerie of variables and changing circumstances that operate within generally unpredictable environments, are a bit more challenging to be automated in their entirety, at least by the end of this decade. This includes functions of skilled trades such as electricians, plumbers, landscapers, childcare providers, etc. Since these are lower wage trades, the Return on Investment (ROI) price tag to automate these trades sometimes is inadequate to justify the investment in automation.

A summary of the occupations that will undergo significant shifts by the end of this decade, based on the 2017 McKinsey & Company (McKinsey Global Institute, Jobs lost, jobs gained: What the future of work will mean for jobs, skills, and wages, 2017) found that occupations in creative jobs such as entertainers, artists, designers, and media workers would fare the best through automation, with a projected 85% employment growth in China, 17% in Germany, 58% in India, -4% in Japan, 28% in Mexico, and 8% in the United States of America. Technology professionals unsurprisingly are well positioned as an occupation by the end of the decade. These include computer engineers, computer specialists, etc. with a projected 50% employment growth in China, 59% in Germany, 129% in India, 15% in Japan, 27% in Mexico, and 34% in the United States of America. The teaching profession, including schoolteachers, postsecondary teachers, other education professionals, education support workers, etc. come in third place as an occupation that will realize solid job gains. These gains include a projected 119% employment growth in China, 17% in Germany, 208% in India, -8% in Japan, 37% in Mexico, and 9% in the United States of America.

Leadership roles are the least number of roles in any organization across any industry. These roles which include managers and executives, are not easily automatable, and will realize modest gains in employment, evidenced by a projected 40% employment growth in China, 21% in Germany, 75% in India, flat in Japan, 24% in Mexico, and 15% in the United States of America. Builders, which includes occupations such as building engineers, architects, surveyors, construction workers, installation and repair workers (buildings and infrastructure), and crane and tower operators will see modest growth, this growth being geographically higher. Builder type trades will witness a projected 9% employment growth in China, 13% in Germany, 117% in India, -16% in Japan, 48% in Mexico, and 35% in the United States of America.

Care providers will continue to be in demand – and this demand has expectedly skyrocketed from a post-pandemic lens. Three years before the pandemic, demand for care providers that included doctors, nurses, physician assistants, pharmacists, therapists, health aides, support, childcare workers, social workers, etc. was expected to experience a 122% employment growth in China, 25% in Germany, 242% in India, -1% in Japan, 83% in Mexico, and 30% in the United States of America. Professionals, a category that includes Account managers, engineers, business and financial specialists, lawyers and judges, legal-industry support staff, math specialists, scientists, and academics was projected 26% employment growth in China, 20% in Germany, 46% in India, 2% in Japan, 32% in Mexico, and 11% in the United States of America.

Towards the lower end of the spectrum are jobs where tasks can be easily automatable for a reasonable cost. These are the occupations that are at most risk for disruption and disintermediation by the end of this decade. Looking beyond 2030, it will be vital to keep in mind that the level of automation shall expand to include the jobs that experience expansion as above. Although the jobs that expand by end of this decade might not be entirely eliminated in the next fifty years, it is important to note that these jobs might look nothing like they do today. Technological change will change them into being unrecognizable versions of themselves in the future.

Job losses due to automation are expected in occupations that can be described as office support, including information and record clerks, office-support workers, financial workers (procurement, payroll, etc.), administrative assistants, etc. The projected employment loss or gain in these professions are dependent on geography, but will overall face slow growth or decline by the end of the decade, with projected 14% employment change in China, -15% in Germany, 21% in India, -23% in Japan, 22% in Mexico, and -20% in the United States of America. Occupations defined by the study as predictable physical work, which includes fine-equipment installation and repair workers, protective services, gaming- industry workers, dishwashers, cleaning- equipment operators, food-preparation workers, general mechanics, etc. are projected at -4% employment change in China, -21% in Germany, 15% in India, -33% in Japan, 8% in Mexico, and -31% in the United States of America. Jobs described as customer interaction type occupations, such as food-service workers, sales workers (retail and online), therapeutic workers (personal trainers), entertainment attendants, personal-appearance workers, hotel and travel workers are projected at 36% employment change in China, 11% in Germany, 46% in India, -13% in Japan, 15% in Mexico, and -1% in the United States of America.

Although this occupational category posts positive growth numbers, the growth is markedly slower than in decades past for this sector. Finally, the study concludes that occupations categorized under unpredictable physical work, including specialized mechanics and repair, emergency first responders, material movers and loaders, machinery installation and repair workers, agricultural field workers, transportation maintenance, building and grounds cleaners, etc. will face the most challenging job prospects due to technology and automation. This sector will see projected employment change of 12% in China, -6% in Germany, 9% in India, -5% in Japan, 16% in Mexico, and 6% in the United States of America. Just like with customer interaction type jobs, despite anemic growth, these sectors have typically been much more robust in job creation in the past decades.

This paper detailed jobs with tasks that can be easily automatable and ergo shall see the greatest disruptions by 2030, as well as occupations that are relatively inoculated from any major job reductions in this time period. However, the commonality here is that both categories of occupations –

those that will see erosion by 2030, and those that will shift in tenor significantly over the next fifty years – will require each of us, the occupants of these occupations, to think and do differently than today. Our **thurvival** will depend on our ability to adopt and adapt rapidly.

A McKinsey & Company 2017 report, three years before the COVID-19 pandemic, took stock of digitization across industries and declared that "Automation will have a far-reaching impact on the global workforce." (McKinsey Global Institute, Jobs lost, jobs gained: What the future of work will mean for jobs, skills, and wages, 2017). The research found that as of 2017, approximately 50% of current work activities are technically automatable by adapting currently demonstrated technology, and that 60% of current occupations have more than 30% of activities that are technically automatable.

The study declared that "The results reveal a rich mosaic of potential shifts in occupations in the years ahead, with important implications for workforce skills and wages. Our key finding is that while there may be enough work to maintain full employment to 2030 under most scenarios, the transitions will be very challenging—matching or even exceeding the scale of shifts out of agriculture and manufacturing we have seen in the past." (McKinsey Global Institute, Jobs lost, jobs gained: What the future of work will mean for jobs, skills, and wages, 2017). According to this study, the percentage of workers in terms of FTE (full time equivalent) potentially displaced by adoption of automation was 10 million if automation was slowest, 15% (equating to 400 million workers) at the speed of automation midpoint, and 30% (equating to 800 million) at the fastest. Based on the pandemic, we are likely closer to the faster end of the spectrum of automation than at the midpoint. The study stated that in the fastest automation adoption scenario, 14%, or 375 million workers could need to change their occupational category. In the workforce of 2030, equating to

2.6 billion people, nearly 9% of global workers will be in new occupations that do not exist today.

This same McKinsey study (McKinsey Global Institute, Jobs lost, jobs gained: What the future of work will mean for jobs, skills, and wages, 2017) also projected a list of existing occupations that would see the most number of opportunities gained and those jobs that would have a net loss of positions. This 2017 study revealed that there are geographic disparities for which occupations loose or gain how many positions by the end of this decade. The study assumed a midpoint level of automation (automation that would displace 400 million workers) – an interesting data point to keep in mind since this was from a pre-pandemic perspective, the pandemic having hastened digitization across industries.

Opportunities Will Abound Outside Our Imagination

Over the next fifty years, the rapid pace of technological innovation, automation, and the changing demands of society will create many new sectors, industries, and consequently - professions - that simply do not exist today. It will be challenging for us to envision and articulate an exhaustive list of such emerging opportunities over the next five decades. This phenomenon is similar to when the Apple iPhone spawned off an entire ecosystem of industries – some that we could easily envision (such as cell phone case manufacturing), and others that we could not – such as the app-based economy. Go back further in time, and one can point to the commercialization and ubiquity of electricity as transformative to society – spinning off countless other industries from it (apart from the obvious light bulb manufacturing industry). Go back even further in time and one can point to the invention of flight, the automobile, the mechanization of industries, etc. as the catalysts that spawned innumerable new industries, and implicitly, thousands of new professions. Suffice to say that while every Industrial Age has brought with it the demise or decline of some industries, it has also created many more industries for each industry that it has disrupted or disintermediated.

It is important to remember that postulating some of the professions that do not exist today is based on our understanding of the technology that exists today, and extrapolating what kinds of jobs that the evolution of this technology might lead to. In other words, any possible professions that you can imagine at this moment in time are a projection into the future within the confines of what we know as of today.

It is unlikely that anyone can predict the kinds of jobs that might be looking for skilled workers in the next 50 years with precision. For instance, the job of "electrician" was not a concept anyone could have predicted in 1900, nor would anyone have been able to envision any industries that have resulted due to electricity – everything from home appliances manufacturing to electric vehicle maker - and all types of jobs these industries have created. The full range of possibilities - industries and professions of tomorrow - in this sense can be quite unimaginable. It is safe to assume that if you presume that the next fifty years would lead to the creation of, say, 100 new industries and 1000 new professions within these industries, the actual number might quite likely be 10x this number or higher.

Despite the fact that it is impossible to predict the full range of opportunities that will present themselves over the next 50 years, it can be a good exercise to think of some of the more plausible industries and/or professional possibilities that might likely emerge over the next few decades. The list of industries and professions below are a minute microcosm of the kinds of opportunities that we might discover in the next few decades.

List of Potentially New or Expanded Industries

1. Space tourism and exploration.
2. Personalized medicine and healthcare.
3. Sustainable energy and clean technology.
4. New and unforeseen Artificial intelligence (AI) and machine learning industry offshoots.
5. Virtual and augmented reality.
6. Quantum computing and cryptography.
7. Advanced robotics and automation.
8. Biotechnology and genetic engineering.
9. Autonomous vehicles and transportation.
10. Drone and unmanned aerial vehicle (UAV).
11. Smart home and Internet of Things (IoT).
12. Renewable energy and energy storage.
13. Personal robotics and companion.
14. Space mining and asteroid resource extraction.
15. Nanotechnology and nanomaterials.
16. Food technology and alternative protein.
17. Wearable technology and smart textiles.
18. Urban farming and vertical agriculture.
19. Carbon capture and climate change mitigation.
20. 3D printing and additive manufacturing.
21. Virtual reality gaming and esports.
22. Advanced materials and graphene.
23. Water purification and desalination.
24. Artificial intelligence in agriculture.
25. Healthtech and remote healthcare.
26. Space-based internet and communication.
27. Sustainable fashion and circular economy.
28. Personal data management and privacy.
29. Neurotechnology and brain-computer interfaces.
30. Autonomous delivery and logistics.
31. Cognitive computing and brain-inspired computing.
32. Biometric technology and identity verification.

33. Renewable resource management.
34. Smart cities and urban planning.
35. Advanced manufacturing and industrial automation.
36. Gene editing and gene therapy.
37. Cognitive enhancement and nootropic.
38. Personal finance and wealth management.
39. Human augmentation and bionic.
40. Space-based solar power.
41. Personalized nutrition and dietary planning.
42. Emotional AI and affective computing.

List of Potentially New Professions

1. AI Ethicist - AI, as an exponentially growing field of technology, will lead to the establishment of a plethora of job opportunities and spawn countless new industries. Technology professionals with expertise in programming and engineering will be in high demand to create and manage advanced AI systems. Some of the resulting professions within these industries are quite obvious and straightforward – AI model developers/programmers, data scientists, data analysts, quality assurance, technology infrastructure, cybersecurity professionals, and so on. Additionally, there may be a growing need for professionals specializing in AI ethics and governance, to ensure that these systems are being developed and used in responsible and ethical ways. With most enterprises having codified ethics and conformity to the highest standards of ethical behavior in their mission and values, it is foreseeable that AI Ethics might be a field that offers opportunities in the near future. Due to the focus on the ethics of AI, the role of "Artificial Intelligence Ethicists" could likely become a critical role in the future. With the certain ubiquity of AI in a wide range of industries, these professionals will be responsible for ensuring that AI is used in a responsible and ethical manner, and for mitigating any negative consequences that may arise from its deployment.

2. Virtual Experience Designers - who will be responsible for creating and designing immersive virtual experiences for consumers. With the growth of virtual reality and other emerging technologies, the demand for skilled professionals in this field is expected to skyrocket. These professions will specialize in creating and designing immersive virtual experiences for entertainment, education, and other purposes. They will use cutting-edge technology such as VR and AR to create interactive, multi-sensory environments that will allow people to engage with digital content in new and exciting ways.

3. Cyber Psychologists - who will specialize in understanding the psychological impact of technology on individuals and society as a whole. As technology becomes increasingly integrated into our daily lives, the need for professionals who can help us navigate its psychological effects will only continue to grow.

4. Renewable Energy Technicians - These professionals will be responsible for the installation, maintenance, and repair of renewable energy systems, such as solar panels and wind turbines, ensuring that clean energy remains a critical part of our energy mix for generations to come. This field is expected to see significant growth is renewable energy and sustainability. As the world becomes increasingly focused on reducing its carbon footprint and mitigating the effects of climate change, professionals with expertise in renewable energy technologies and sustainable practices will be in high demand. This could include roles such as renewable energy engineers, sustainable city planners, and carbon offset specialists.

5. Biotech - in the field of biotechnology, there may be a growing demand for professionals in the areas of gene editing and synthetic biology. These professionals will play a crucial role in developing and managing cutting-edge technologies that have the potential to cure diseases and enhance human biology in unprecedented ways.

6. Cultural Ambassadors - as the world becomes increasingly interconnected and interdependent, there may be a growing need for professionals with expertise in cross-cultural communication and global governance. These individuals will be instrumental in fostering cooperation and understanding between nations and cultures, helping to create a more peaceful and prosperous world.

7. Space Settlement Architects - another new profession that may emerge is that of space settlement architects, who will design and plan self-sustaining communities on the Moon, Mars, and other celestial bodies. They will need to have a deep understanding of the unique challenges of living and working in space, and will be tasked with creating living and working environments that meet the needs of humans.

8. Digital Detox Therapist: Helping individuals manage and overcome technology addiction.

9. Space Tourism Guide: Assisting travelers in planning and experiencing trips to space.

10. Drone Traffic Controller: Regulating and managing the traffic of autonomous drones.

11. Genetic Counselor: Providing guidance and support for individuals making decisions based on genetic information.

12. Sustainable Energy Consultant: Advising individuals and organizations on adopting renewable energy solutions.

13. Blockchain Auditor: Verifying and ensuring the integrity of blockchain-based systems and transactions.

14. Data Privacy Manager: Protecting personal and sensitive data in an increasingly connected world.

15. Robot Mechanic: Repairing and maintaining advanced robots and robotic systems.

16. Augmented Reality Designer: Creating interactive and immersive augmented reality experiences.

17. Urban Farmer: Cultivating crops and managing food production in urban environments.

18. Personalized Medicine Consultant: Assisting individuals in customizing medical treatments based on their unique genetic makeup.

19. Cybersecurity Analyst: Protecting organizations from cyber threats and hacking attempts.

20. Quantum Computing Specialist: Developing and utilizing quantum computing technologies.

21. Drone Delivery Specialist: Managing and optimizing drone delivery operations.

22. Personal Branding Consultant: Assisting individuals in building and managing their personal brands online.

23. Climate Change Analyst: Assessing and mitigating the impact of climate change on businesses and communities.

24. 3D Printing Engineer: Designing and optimizing 3D printing processes and materials.

25. Telemedicine Specialist: Providing medical consultations and diagnoses remotely.

26. Autonomous Vehicle Designer: Creating vehicles and systems for self-driving transportation.

27. Remote Work Facilitator: Helping organizations and individuals optimize remote work setups and productivity.

28. Virtual Reality Fitness Trainer: Designing and leading virtual fitness programs and workouts.

29. AI-Assisted Healthcare Technician: Operating and maintaining advanced AI-powered medical equipment.

30. Neuro-Implant Technician: Installing and maintaining neural implants for various applications.

31. Drone Journalist: Reporting news and events using drone-based footage and reporting techniques.

32. Virtual Reality Therapist: Conducting therapy sessions using virtual reality environments.

33. Cybernetic Organ Designer: Developing and implanting advanced artificial organs.

34. Space Debris Engineer: Designing solutions to mitigate and clean up space debris.

35. Drone Cinematographer: Capturing aerial footage and cinematography using drones.

36. Mindfulness Coach: Guiding individuals and organizations in practicing mindfulness techniques.

37. Nanobot Engineer: Designing and developing nanoscale robots for medical and technological applications.

38. Drone Swarm Coordinator: Managing coordinated movements and operations of drone swarms.

39. AI Personality Designer: Creating unique and personalized artificial intelligence personalities.

40. Climate Change Adaptation Planner: Developing strategies to help communities adapt to the effects of climate change.

41. Virtual Reality Travel Agent: Planning and organizing virtual travel experiences and destinations.

42. Human-Machine Teaming Specialist: Facilitating collaboration and synergy between humans and intelligent machines.

43. Asteroid Miner: Extracting valuable resources from asteroids and celestial bodies.

44. Robotic Companion Developer: Designing and programming interactive robotic companions for various purposes.

45. Genetic Engineer: Manipulating and modifying genetic material for medical and agricultural advancements.

46. Space Traffic Controller: Regulating and managing the movement of spacecraft in Earth's orbit.

47. Digital Legacy Manager: Handling and preserving individuals' digital assets and online identities after their death.

Survive. Thrive. Thurvive

While we cannot predict all the industries and professions that might emerge with a level of precision, what we can certainly do is to prepare ourselves to capitalize on these opportunities. The tips and tricks to thrive amidst technological change are transferrable whether these future industries are "greenfield" industries, or technology-driven evolutions of existing ones. This is simply based on the fact that "thurvival" is not about preparing for a specific profession in a specific industry, but about preparing yourself to capitalize on technological change, regardless of profession or industry.

We have all been spectators to the sport of large-scale change; both, in our personal and professional lives. However, there will be no mere spectators of this sport over the next twenty-five years. We will all - each of us - need to be active participants because a large number of professions will feel varying degrees of disruption over the next twenty-five years, and most professions will be disrupted, some to the point of being unrecognizably transformed, over the next fifty. Said another way, there will likely be no job – no profession – across any company in any sector of any industry that will be untouched and unchanged by technological change over the next several decades. We tend to delude ourselves by lulling ourselves into a false sense of security with the illusion that any change will happen to someone else, somewhere else, and in some industry. We tend to develop a cultural complacency to change. Those of us who do not prepare ourselves to rise above our own "cultural change complacency" in this "Fourth Industrial Revolution" will likely be shaken out of our stupor by the fierce, unrelenting, and rapid pace at which change will happen. Unless we invest the time in shifting our mindset now, we will be forced to change under duress.

We were introduced to the concept of "**thurvival**" at the outset of this book. Through our journey thus far, we have acquainted ourselves with the breakneck pace of change that is all around us and exponentially accelerating. We examined our own "cultural change complacency" – the inertia that prevents us as people from managing change we perceive as outside our control, and being able to adopt and adapt in time to stay abreast, or better, get out ahead of the technological change. We also examined our own "corporate change complacency" – the inertia that prevents organizations from foreseeing change, and being able to adopt and adapt in time to stay abreast, or better, get out ahead of the technological change. We understood the reason why we tend to resist change in the workplace, while being amenable to technological change at home and for things we believe we can exercise control over.

We reviewed example after example of successful corporations that failed to anticipate change and change themselves in time. We delved into the fact that no job in any sector in any industry will be shielded by change over the next twenty-five years. While the next several decades will create brand new jobs in brand new sectors in brand new industries, it will also mean outright elimination of entire occupations, while leaving no occupation untouched by change, and rendering some completely unrecognizable from what they look like today. In order to ensure each of us, whether we are 25 or 52 years old, have a long, stellar career that we retire from under our own volition, we need to survive our own instincts against change, our own change resistance hardwired paradigms, and set ourselves up to thrive in the new economies of the 21st century.

Learning the tools for **thurvival** in the "Fourth Industrial Revolution" is a necessity, not a luxury. Each of us need to be thoroughly prepared and equipped to meet the challenges and capitalize on the opportunities that await us over the next twenty-five years in the midst of accelerating and exponentially growing technological change. While it is alluring to fantasize that your own job or your own organization and industry might be unaffected, or minimally impacted, by this accelerating technological change over the next twenty-five years, I can almost certainly assure you that there will be no sector or any industry that will be spared the disruption of technological change. The only way to ensure that your own job and your own profession can survive and thrive amidst these changes is to recognize that this is happening, that it is inevitable, and nothing you can do can stop it. You can only prepare yourself, your friends and family, and your organization to anticipate and be ready to meet the moment.

As we conclude the journey to ground ourselves in the "today" of where we are as a society dealing with exponential and accelerating technological change, our understanding of how to orient ourselves for sustained success in the future is only beginning. The remaining chapters of this book will prescribe actions we can take to ***survive*** our own reluctance to change, and equip us with some of the techniques in order to ***thrive***, **thurvive**, in the "Fourth Industrial Revolution". It is important that you should be prepared to commence implementing the techniques prescribed here as soon as you practically can. These generally turnkey recommendations are actionable and practical things that, with some planning and organization, can you can execute upon without any cost or without any significant upfront effort. Note that just as technologically will continue to change in an accelerated and exponential rate, these "**thurvival**" techniques are intended to develop a lifetime of discipline. There is no start and stop, but instead reprogramming ourselves to be able to swiftly adopt and adapt, and that reprogramming should commence immediately. Once again, as President

John F. Kennedy said, "the best time to repair the roof is when the sun is shining".

Chapter Fifteen - Thurvive

Autonomous flying cars, robots automating everyday time-consuming tasks, connected everything, AI ubiquitous across every facet of our lives, the predicted "Singularity", and more! Imagine for a second the occupations, industries, ecosystems, and jobs that these and many other technological advancements are going to foster and nurture. We are heading into an exciting time, one that offers unlimited opportunities. These opportunities will only be available for those who are able to capitalize on them. Being able to capitalize on these opportunities will require us to anticipate them, expect them, and prepare for them.

Consider the plethora of new opportunities that entire industries that are being created all around us today. It was only in 2012 when the first generally accepted electronic vehicle (EV) was introduced, the Tesla Model S. Having been deemed as a viable business, Tesla Motors served as a catalyst towards the EV movement, despite electronic cars having existed in some form for long before. Most major automobile manufacturers are promoting their electric fleets over their traditional combustion engine ones. Several cities and some countries have plans to outright ban the sale of non-EVs within the next decade. From the ongoing maturity of solar power to store and distribute plentiful, renewable energy – to charge everything from EVs to power homes, to the super charges being developed across the United States for rapidly charging these EVs in a manner as ubiquitous as the availability of gasoline, think about the fact that none of these sectors of the economy existed over a decade ago, and if they did, they were considered as specialized and niche. Think of the number of new jobs and the new types of employment and occupations created by this growing industry. Think about the ecosystems it has created and the countless many industries it has spawned.

Robots have been deployed in Japan for elder care and to address the needs of that country's aging population for a few years now. Robots deliver food to college kids across university campuses in the United States today. Robots serve as food delivery, effectively being waiters at restaurants such as Buffalo Wild Wings in the United States. Intentionally friendly-looking robots operate at Dubai International Airport for screening passengers for symptoms and markers of COVID-19. Robots, once considered as a novelty less than five years ago will with certainty become an integral part of industries and our daily lives globally over the next decade. The widespread use of drones is transforming everything from Amazon package deliveries and food deliveries using popular delivery services such as UberEats and GrubHub to how wars are conducted.

AI is on the precipice of a milestone, where Artificial General Intelligence (AGI) will find itself disperse throughout our everyday technologies such as with OpenAI's ChatGPT imminently finding itself being embedded across Microsoft Office products (Metz & Weise, 2023), and competing against Google as a viable search engine. Think about the professions this will transform – from being able to automate the authoring of legal briefs to generating entire books (or book forewords like for this particular book) – AGI will create as many more abundant new occupations than it disintermediates.

We tend to look at disruptions that innovations cause. It is where we naturally gravitate. It is alluring to, in retrospect, consider the decisions made by firms that were disrupted and industries that are now defunct. It is intriguing to read about case studies of disintermediation due to innovation. However, innovation progresses us as a society and creates many more opportunities than it disrupts. Consider the iPhone. Innovations such as the iPhone are fundamentally transformative and their effects are felt across sectors and industries. While these innovations can be disruptive to some sectors, leaving entire industries defunct in their wake, the ripple effects they create reverberate across societies, cultures, occupations, and economies.

The iPhone for instance isn't just a windfall for Apple. The industries it has created around it – from apps that can generate tens of millions in revenue and spawn franchises (just think of the game, Angry Birds), to iPhone phone case manufacturers, docks and Bluetooth speakers, and on and on – have provided new job opportunities for millions of workers all around the world. In addition to leaving Nokia defunct in its wake, the iPhone can also be directly attributed to everything from the death of the CD industry, the CD player industry, suppliers of materials to those industries, all the way to the firms that used to manufacture the printed CD sleeves and covers. Indirectly, one could establish causality from the iPhone to the downfall or downright demise of a multitude of industries – from cable TV (streaming apps), Skype (replaced by FaceTime), DVDs and Blu-ray media (and the downstream impacts akin to CDs), and countless others.

The following "**thurvival** tale" is comprised of a pair of two real-life examples, both which I have had personal experiences with. The first story, replicated from "Find Your Red Cape", my book on leadership, focuses on an individual who – despite being an exceptionally capable and experienced professional – fell short of evolving their career to adopt technology and continually grow themselves. The second story is diametrically in contrast to the first individual, and is a real example of someone who invested herself in the continual pursuit of adopting and adapting with changing times. She kept herself and her profession vibrant, and as importantly, assured herself of continued gainful employment in an industry that very easily automated the job that she started her career with. Challenging herself to continually grow and adapt to changing circumstances, learning new tools, and learning to incorporate the latest technological advancements into her job, she continued to reinvent her role through her years of service. In being able to deliver business value despite technological advancements around how her firm conducted business, she ensured that she always was in command of her professional future and is a remarkable case study on how to become an effective steward of your own career.

A Thurvival Tale

1. Story 1 - Being static is not the same as being stable

"I regret to inform you that your position has been eliminated," I repeated that statement multiple times over that day to individuals across my teams. I had received a list of "impacted" employees from Human Resources (HR) that somber morning. On that list were names of individuals whose employment with the company was being terminated due to corporate restructuring initiatives. As the senior division leader, with a few employees in leadership roles themselves being affected, it was my responsibility to carry out conversations with those impacted. While this was not entirely unanticipated due to the massive digitization-fueled corporate restructuring, it was quite unexpected – especially since I was only a few months into my tenure in this senior position. Senior leaders are typically brought into reviewing and vetting lists months in advance with HR and other stakeholders in utmost confidence and secrecy to provide guidance and difficult recommendations. Most of the time with large-scale shifts and impacts of this nature, most organizations require even their most senior management teams to sign Non-Disclosure Agreements (NDAs). As soon I was handed a list and a "script" from my Human Resources partners, I know that this was not one of those instances. While it was a bright and beautiful Spring day outside, the mood inside the building could not have been darker and grim.

The most difficult part of this difficult day was to inform one of the leaders on my team, that his position was being eliminated. I held deep admiration and respect for this individual and his four decades of experience. He had started working with this company almost a day to the month and year that I was born. In fact, that particularly difficult day happened to coincidentally be my 40th birthday, and he was ten days shy of his 40th anniversary with the company. He knew the business better than most, but had not kept up in learning new technologies or evaluated new ways of doing things. Perhaps his inability to continually reinvent himself and explore how to incorporate technology into his own job and combine that with the might of his experience came from inertia, fear of the unknown, or complacency due to being in a comfort zone. Getting close to retirement age, maybe he didn't think he had to.

Understandably, a few years shy of retirement, this unexpected termination of his employment usurped the retirement plans he had set with his spouse. Unraveling their household plans for savings and for post-retirement travel and adventures on a fixed budget, the layoff – despite the generous severance package – upended plans that had been nurtured for several years. Putting aside the sheer shock of being let go from an organization that you have invested over four decades of service to, this one event belied the significant experience that the individual had accrued over these four decades. This job and this organization was all that this person knew. Every major event in his life had occurred as an employee of this organization – weddings, including his own, birth of his children, graduations, birth of his grandchildren, birthdays, memorials, anniversaries, and so on. Everything about him – from how he paid his utilities to his very identity – had been wrapped around his job within this global behemoth of an organization.

In retrospect, reflecting on the day, I can understand why his name would have appeared on the list. Whereas I valued his experience and was able to marry his keen business insights with technologically progressive ideas from his peers, prima facie – on face value – for anyone who would not have known him and his contributions, his business value would not be apparent. Without having been involved in the decision to eliminate his position, I can see that a third party consultant would have evaluated his job responsibilities, conducted some analysis, and deduced that he was performing similar tasks and functions that he had been a decade prior to his layoff. Even going back two decades, or three, the general parameters of what he did and how he did it would have been just as recognizable. There was likely a sense that if a person is doing the same exact function that they performed a decade prior, they have likely not kept up with the times and their role can either be automated significantly, or they are not the type of individual that the firm wants to help shepherd the organization into the Fourth Industrial Revolution. In any scenario, regardless of the thought process that went into the role elimination, the rationale for the choice of that specific person's role being eliminated was quite apparent – here was a long tenured employee with exceptional experience accrued over decades, albeit doing the same job that he was a decade or two decades prior. These aren't the skills they needed to propel themselves forward.

I think of this individual quite often. As much as I respected his wealth of knowledge and valued him as a member of the team, I can see the rationale behind why his name appeared on a list. And although he has done well for himself in retirement, I cannot for the life of me imagine the stress and the upheaval that this incident created in his life – and that of his family. A few years shy of retirement with defined plans and goals for his post-retirement life, the angst that they must have felt to adjust these plans under duress is unfathomable to me. I found myself speculating that his story could very well be the story of countless others, and it was an impetus for me to deliver a TED Talk on the subject and invest the time in writing this book. If this clarion call and influence even one person – a single reader – to look at their career amidst the age of technology differently, and take action to proactively act differently, I would consider that a win.

2. Story 2 – Continual learning in action

As gut wrenching as the first example is, contrast it with a story of the mother of a close friend. I met her a few years ago when I visited him in New York City. Living only a few blocks from him, during our brief conversation over coffee in the Upper West Side, she told me about her new job as a "Lead Technology Enabler." "What does your work involve?" I had asked her cautiously because I've always known her to be a clothing template designer, a seamstress by trade. Had she suddenly switched careers? Gone back to school and moved from apparel design and sewing at her job in the Garment District to software programming? Nope. Unlike the colleague I'd had to let go, she had been just brilliant at keeping up. She had, through her career, demonstrated an exceptional and intentional ability at adapting to evolving technology within her profession and coopting this progress to better her job and that of her firm.

When she started her career 45 years ago, it was all manual - designing apparel using markers, measuring tape and scissors. Then came machining and she taught herself how to augment her job with machines. Next it was computer-aided design, so she taught herself how to perform the same tasks on the computer and let the machines do the rest.

A few years ago, she became the mentor to younger employees coming into the industry because they could do the work on the computer but lacked her practical experience to discern flaws in the templates. She told me that she expects to be learning Artificial Intelligence algorithms before she finally retires. As you can imagine, I was on the edge of my seat to learn more. She proceeded to take me on a fascinating journey. "Imagine walking into Men's Warehouse tomorrow", she urged me. "You are presented with an iPad of suits to choose from. You select a suit and walk in front of Artificial Intelligent cameras, like the set up to scan you in the airport. The AI algorithms detect your size, weight, gait, height, body type, etc. and customize a suit for you. Totally free of needing to interact with anyone, and certainly free of intrusive tailoring sessions." When I shared this story with a female friend her reaction was "Can they do the same thing for bathing suits? I don't think any of us will miss dressing rooms and that oh-so-flattering fluorescent lighting."

What is the difference between my friend's mother and the co-worker I had to let go? Both were in a similar age group. Both were smart and capable. But one grew her skills, adopted and adapted to technology, reinventing herself along the way. My co-worker made only incremental progress within what he knew best. The premise of this pair of stories is the central premise behind this book - to encourage you to future enable your OWN career - whether you are 22 or 62 - so that the story of the person whose employment was disrupted doesn't become your story.

We're going to explore how to do just that.

A Recap

The next several decades are poised to witness an unprecedented wave of technological change, spearheaded by technology in general, and specifically by artificial intelligence. As AI continues to evolve and permeate various aspects of our lives, it offers a plethora of opportunities for humans to thrive. However, it is crucial for us to understand the challenges that lie ahead and the necessary steps we must take to adapt, survive, and flourish in this new era. We will further explore the exciting possibilities that await us and discuss the essential strategies for navigating the transformative landscape of AI-driven changes.

We know for certain that our jobs will change. What we don't know is when specifically this tsunami of change will wash over our specific industry, our specific firm, and our specific job. We know for a fact that none of us will be successful at rigidly preserving our occupations just the way they are. What we also know for a fact is that each of us need to commence preparing for this future. How does anyone - Generation X, Millennial, or Generation Z go about preparing for this future? A recap of our journey thus far will be helpful before delving into **thurvival** techniques in order to achieve sustained success in this exponentially accelerating digital world.

Here are twenty things that we know:

1. There will be unprecedented technology-based disruption overs the next two to three decades.
2. These technology-based changes will accelerate exponentially.
3. Industries and companies that have been slow to digitize will struggle.
4. Individuals who fail to recognize the opportunities and the challenges posed by these changes will struggle.
5. There will be no facet of our lives that will be impervious to technology. Technology will influence and be increasingly incorporated into our personal and professional lives.
6. It is a near certainty that most occupations in the next few decades will be unrecognizable to us today. Existing occupations will look nothing like they do today.
7. Technology, including AI, will allow for massive automation over the next several years. This level of AI-driven automation will continue to increase and mature rapidly to become increasingly more sophisticated.
8. This level of automation will focus on occupations that are heavy on repeatable, operational tasks with predictable variables and predictable outcomes.
9. These occupations, typically reliant on lower-skilled labor, will be completely eliminated.
10. Existing occupations, regardless of level of skill required, will all require skilling and reskilling.
11. Even the occupations that cannot easily be automated, will leverage technology to varying degrees. This will therefore require occupants of

these professions to reskill themselves to augment aspects of their jobs with technology.

12. The "Fourth Industrial Revolution" will create hundreds of brand new industries that do not exist today, nor can we easily envision.
13. These hundreds of brand new industries themselves will spawn hundreds of other new industries in a compounding effect.
14. This compounding effect will establish entirely new industry ecosystems.
15. These new industries and new ecosystems will create countless new occupations and these new occupations will create millions of new jobs.
16. Humans are generally change-embracing in their personal technological lives, but generally change-averse in their professional technological lives. This embrace and aversion dichotomy is simply because we believe to be in charge and in control of our personal technological choices, while believe that change is done to us – in the absence of choice – in our professional lives.
17. We have to survive our own instincts against embracing technological change at work. We need to survive the natural reticence of changes to the jobs we know. We need to survive our tendency to settle into a corporate comfort zone.
18. We have rethink our relationship with technology in the workplace. We have to rewire our thinking about change in order to ensure our careers are readied to withstand the onslaught of technological change.
19. With surviving our own instincts as a foundation, we need to take actions in order to capitalize on the opportunities that will be abound in this digital era. We need to enact actions to thrive amidst this accelerating change.
20. Taking action in order to survive and thrive cannot wait. We need to plan, prepare, and execute on these actions starting now. Since we've established that the future is already here, and arguably that the *future was yesterday*, it is all the more imperative to start immediately practicing "**thurvival**" techniques.

How To Thurvive

As we discussed at the start of this book, **thurvival** will require us to do the following:

i. Understand the change - The first step is to recognize that the technology-driven changes are an inevitability. We took a step into the future in the first section of this book and took a peek at the most conservative likely look into our future. There will more than likely be technology introduced over the next few decades that is too fantastical to envision today.

ii. Understand the need to have to change – The second step is to understand that we do not have a choice to slow, stymie, or stop these technology-driven changes. In section two of this book, we explored the past and sought to ground ourselves on the fact that while the past offers some prologue to the future, the increasing and accelerating pace of change over the past century is a harbinger on shape of things to come.

iii. Understand the need to want to change – The third step is that we ensure we are proactively addressing change readiness. In section three, we came back to the present after having explored the future and peeked at the past. We examined how the recent past is influencing our world today and what that might portend for our occupations over the next few decades.

iv. Understand how to be ready for the change – The fourth step is to adopt the recommendations outlined in subsequent chapters, building your own personal plan on the foundation of a continual learning mindset and keep exercising that continual learning mindset. The following are the five Thurvival Techniques that we will delve into over the next few chapters. These techniques are a set of four recommendations that will help you survive the coming digital disruptions and thrive amidst the countless opportunities that these digital disruptions will generate.

Thurvival Techniques

The following four core techniques for "**Thurvival**" will be examined in further detail in subsequent chapters. Following these tenets will help you in the journey to "future proof" your career, and be better prepared to survive and thrive – "**thurvive**" the "Fourth Industrial Revolution".

Thurvival Technique 1. Rewire your thinking – Learn, Unlearn, Relearn (Chapter Sixteen)

Thurvival Technique 2. Harness Technology (Chapter Seventeen)

Thurvival Technique 3. Understand your Industry and Company Trends (Chapter Eighteen)

Thurvival Technique 4. Develop your Human Skills (Chapter Nineteen)

The next twenty-five years offer us a future that is full of possibilities. It is a bright future if we can proactively prepare to greet it and seize the opportunities that it brings. This future can seem daunting because the speed with which technology is changing can make the journey seem uncertain. However, creating a roadmap on how we can traverse this landscape will yield us certain success, regardless of the destination. There is a vibrant future that we have the opportunity to build with the choices that we make today, and these are the choices that we will put forth as recommendations in the remainder of this book. All this change – the pace, the acceleration, the transformations, the disruptions – can all feel overwhelming. It can feel uncomfortable. Change can seem daunting, but it also brings abundant and unlimited opportunities and possibilities to thrive. Learning to "**thurvive**" will help us to see how we can embrace change as an opportunity for growth and progress, not as a threat to our comfort. The four core "**thurvival**" tenets will provide you with techniques to be able to do just that.

Chapter Sixteen – Thurvival Technique 1: Rewire Your Thinking: Learn, Unlearn, Relearn

Chapter Seven of this book opened with an examination of two frequently cited quotes about the ability of species to survive by being adaptable to change in relation to changes in their external environment. "It is not the strongest nor the most intelligent of species that survives, but the one that is most adaptable to change." (Darwin, 1859) – Charles Darwin. "It is not the most intellectual of the species that survives; it is not the strongest that survives; but the species that survives is the one that is able best to adapt and adjust to the changing environment in which it finds itself" (Megginson, 1963) – Dr. Leon C. Megginson.

The chapter delved deeper into the concept of change as relates to a natural resistance to change that most individuals exhibit when presented with change that they believe they have little control and governance over. The ability for any species – specifically for us as humans in the workplace – to survive, and even thrive, amidst a rapidly changing environment will always come down to how quickly we can adjust and adapt to this changing environment. At the very core, the ability to adjust and adapt within the context of the workplace will depend on our ability to continually learn. Being a lifelong learner and embracing a growth mindset isn't only about continually learning new things through a myriad of tools and sources available to us. When it comes to lifelong learning, the key is a shift in our own mindset, and this mindset shift is a fundamental difference in how we should think of approaching continual learning going forward. This rewiring of your thinking is a term that we broached earlier in this book. Popularized by American futurist, Alvin Toffler, this rewiring of our thinking is best characterized as our ability (and need) to be able to "***learn, unlearn, and relearn***".

Learn, Unlearn, and Relearn

The ability for us to "Learn, Unlearn, and Relearn" faster than we have ever done before will prove to be decisive in maximizing our potential to thrive in the Fourth Industrial Revolution. Learning, unlearning, and relearning are not mutually exclusive, but are equal parts of a continual and closed loop. Each of us will be required to learn x, unlearn y, and relearn z at times, and learn x, unlearn x, and relearn y at others – often having to juggle several of these learn-unlearn-relearn sets concurrently. So what does it mean to "learn, unlearn, and relearn"?

To ***learn*** means to ensure we continually pursue new knowledge, learning new ways of doing things, and being receptive to change even when it feels uncomfortable, and sometimes, especially when it feels uncomfortable. To ***unlearn*** means that we need to rapidly be able to shed our old habits and ingrained paradigms, and being open to new tools and processes without being mired in the "this is the way I've always done it" ideology. To ***relearn*** means to know that everything that we have just learned is likely to be usurped by a newer and better way of doing things. To ***relearn***, we need to expect that even though we might have just learned a new skill, it will be outdated much faster than ever before, and we will need to keep up. This, in a nutshell, is the central predicate for rewiring our thinking – adopt and adapt by being able to ***learn, unlearn, and relearn***.

In the Fourth Industrial Revolution, the need to rapidly ***learn, unlearn, and relearn*** has become increasingly crucial.

As we have been exploring, the world is evolving at an unprecedented pace, with constant advancements in technology and innovations reshaping industries, transforming job requirements, and altering the very nature of our daily lives. Rapid learning is no longer a luxury, but a necessity. New technologies emerge, trends change, information becomes outdated at an alarming rate, and existing technologies evolve rapidly, creating an ever-changing landscape. To keep up, we must be willing to engage in continuous learning, acquiring new knowledge and skills to stay relevant. The ability to **learn** quickly, absorb information efficiently, and apply it effectively has become a valuable asset in navigating the digital age.

First and foremost, the digital age brings forth advancements and innovations. New technologies emerge. To keep up with these changes, individuals must be open to learning new skills, acquiring fresh knowledge, and staying up-to-date with the latest developments. The ability to rapidly learn allows individuals to remain relevant and competitive in their respective fields, enabling them to seize new opportunities and adapt to changing circumstances.

Equally important is the concept of **unlearning**. As technology advances and paradigms shift, outdated knowledge and practices can become barriers to progress. Unlearning involves letting go of obsolete beliefs, assumptions, and habits, creating space for new ideas and perspectives to take root. It requires us to challenge our preconceived notions, question established norms, and embrace the discomfort of change. Unlearning paves the way for fresh insights and innovative approaches that align with the demands of the digital era.

Relearning completes this triad of adaptability. It involves the ability to continuously update and upgrade one's skills and knowledge base. Relearning acknowledges that what was relevant yesterday may not be relevant tomorrow. It is a proactive response to the ever-evolving technological landscape, ensuring that individuals are equipped to face emerging challenges. Relearning empowers us to stay ahead of the curve, seize new opportunities, and effectively contribute to a rapidly changing world.

The Fourth Industrial Revolution requires us to be open-minded, curious, and willing to explore uncharted territories. Embracing these principles fosters agility, resilience, and adaptability - qualities that are highly sought after in the dynamic job market of tomorrow. This mindset enables us to be active participants in shaping our own futures, rather than being passive observers.

The following three stories are real examples that embody our ability to learn, unlearn, and relearn. The first story, as appeared in my book on leadership, "Find Your Red Cape", is about the Polgar sisters and seeks to highlight the fact that anyone can learn anything and that this learning compounds. The second story is about British table tennis (ping pong) star Desmond Douglas MBE (Member of the Order of the British Empire), and seeks to highlight the fact that unlearning can be just as powerful as learning and learning, perhaps even more so. The third story is about American chess champion, martial arts champion, and author Josh Waitzkin, and seeks to highlight that anyone can relearn anything at any given point in time.

1. Learn – The Polgar Sisters

"Let's get married, and have children as part of a protracted science experiment I am conducting." A proposal framed in the manner, and delivered with the same passion as one would propose an evening of extended calculus problem solving, would likely never be the backdrop to a romance novel, a Shakespearian play, or a Hollywood romantic comedy. And even so, Klara, a Ukrainian foreign language teacher, having found the concept intriguing, said yes. Her suitor was Hungarian researcher, Lazlo Polgar. Lazlo postulated that skilled individuals are created due to repetitive and rigorous training. He believed that anyone can learn anything; and with appropriate training, be an expert. This theory ran generally contrary to the prevailing opinion in the 1960s that people are born with natural abilities that render them adept at certain skills over others. He believed that with regimented training, he could "create" a genius, rather than having one been born with "genius". He believed that training, as opposed to an individual's intellect was the linkage to an individual's accomplishments.

The Polgars were married in 1965, and their first daughter, Susan, was born in 1969. Lazlo decided to commence Susan's training when she was a toddler, picking the strategic game of chess as the crux of his hypothesis. His rationale for choosing chess versus other trained skills was that chess is the only one-on-one game that requires the supposed application of superior strategy, and this strategy is said to be predicated on intellect. Chess, like singles tennis, provided a clear global ranking system that clearly established an individual's superiority, but unlike tennis, was unconstrained by physical ability, and devoid of the risk of physical injury. Chess also offered the Polgars an opportunity to do something tennis could not. Liberated from needing, or having constant access to a tennis court for training, chess allowed them to foster an environment at home where they could surround Susan by all things chess, and amplify her training.

A mediocre player himself, Lazlo provided every opportunity he could for Susan's training on the game of chess. Susan intensely practiced chess on a daily basis through her toddler years, having accrued a significant amount of experience by the age of 5. Lazlo entered Susan in a local chess competition at the age of 5, where she could barely reach the tables where the chess sets were placed. Susan Polgar went onto beat all her adult opponents with a 10-0 score. Her intense chess training continued on, and at the age of 15, Susan was a top-ranked women's chess player. At the age of 17, she qualified for the Men's World Championship and became a Grandmaster in 1991 at the age of 22. Susan is the first woman in history to win the Chess Triple Crown.

Lazlo's hypothesis that training helps forge expertise seemed to have been successfully proven with one test subject, his first born, Susan Polgar. However, Lazlo and Klara were blessed with two other daughters, and the three Polgar sisters definitively underscored Lazlo's theory that skilled experts can be trained for, and that all individuals are born with relatively the same level of "genius".

Sofia Polgar is the second daughter of Lazlo and Klara Polgar. Unlike her older sister Susan, Sofia benefitted from the support structure of three individuals instead of two, amplifying the intensive chess training environment in their household. With her intense chess training, Sofia also became a decorated chess player. In fact, she attained a comparable level of rank and expertise that her old sister Susan did, at a younger age. In addition to winning several medals and tournaments, Sofia was only 14 when she won a chess tournament against several other grandmasters. Her performance at this tournament in Rome put her as the fifth-best player in the history of competitive chess. The level of success that Susan achieved at 15, Sofia attained a measurably comparable level at the age of 14. Lazlo's experiment indicated that intensive training not only develops "genius" level skill, but it actually compounds based on the environment, atmosphere, and support structure.

And then came Judit, proving that not only does intensive training develop more robustly in the right environment, and that it compounds, but it can also compound in a way that it accelerates the achievement of measurable goals. Judit Polgar, the third and youngest daughter of Lazlo and Klara, is widely considered to be the best female chess player in history. Judit not only had her parents providing her the intensive chess training support structure, but now had the expertise of her two older sisters as well. At the age of 9, Judit stunned an awe-struck audience, when, at a tournament of the world's best chess players, she battled, and defeated five opponents concurrently; blindfolded.

Judit Polgar became a Chess Grandmaster at age 15 and is the youngest player to break into the Top 100, something she achieved at the age of 12.

In 2002, Judit beat Gary Kasparov, causing the champion to walk out in angst (hopefully to reflect on the sexist statements he had made in advance of this game). Judit is the only female chess player to win against a reigning World Number 1. No other female chess player has qualified for a World Championship event.

There are several lessons we can derive from the story of the Polgar Sisters and apply them to our learning mindset as we head into the Fourth Industrial Revolution. First, it is evident from the story of the three sisters that anyone can learn any skill at any time. It matters little how young or how old you are. Learning a skill from a very young age – with repeated practice over many years – is unquestionably beneficial, and yet it is only part of the overall story. Consider that of the three sisters, the youngest of them attained the upper echelon in chess at a much younger age than her sisters. The second takeaway from this is that learning compounds. Unlike her older siblings, Judit had the cumulative learning and coaching from her two older sisters, as well as her parents, while the oldest sibling, Susan, benefitted solely from the learning environment that Lazlo and Klara had created for her. Thirdly, the Polgars' experiment proves that anyone can learn anything. While there might be natural innate skills that individuals possess, the fact that the Polgars were able to foster a learning culture where all three of their progeny attained the highest level of success in the same field proves that anyone can learn the same set of skills regardless of "natural born ability" or aptitude. If you believe you're not technologically inclined (or in parlance, "tech challenged"), the story of the Polgar sisters proves that you can learn anything, and learn that anything – including new technology - at any age.

2. Unlearn – Josh Waitzkin

The subject of a movie based on his life, titled "Searching for Bobby Fischer", Josh Waitzkin is an American chess prodigy, marital artist, and author. Waitzkin became interested in the game of chess at a very young age. Born in New York City, Waitzkin would watch people play chess in the park as he would traverse the park with his mother as he passed them when he went to play on the monkey bars. By pure observation alone, Waitzkin was able to learn more about the game of chess and decided to take up the game himself. His competing against other chess players in the park at the age of 6 drew the attention of chess coach, Bruce Pandolfini.

With two years of training under Coach Pandolfini, Josh Waitzkin won the National Primary Championship in 1986 (age 10), the National Junior High Championship in 1988 (age 12), and the National Elementary Championship in 1989 (age 13). Continuing to bolster his chess prodigy credentials, Waitzkin won his national master title when he was only 13, the National Junior High Championship again in 1990, and the National High School Championship in 1991. At age 11, Waitzkin played World Champion Garry Kasparov to a draw. His opponents included the Polgar sisters at various times, and his match record includes a win over Sofia Polgar in 1995. Waitzkin dominated the US Junior chess scene, winning multiple national championships. He became an International Master at age 16, when he was also the co-champion of the U.S. Junior Chess Championship. At 17, Waitzkin won the U.S. Junior Championship and placed in fourth place at the Under-18 World Champions.

It was around that age when Waitzkin, having achieved the top honors at chess, decided that despite all that he had learned in the game, he had reached the limits of what he could accomplish, without broadening his naturally aggressive playing style. Remarkably, Waitzkin stepped away from chess and entered the world of martial arts. Waitzkin went onto win 7 US National medals in Aikido, before turning his attention to Tai Chi, winning the World Championship in Tai Chi Push Hands (the competitive variant of Tai Chi) in 2004 in Taiwan. With a World Championship in Tai Chi, he went onto earn a Black Belt in Brazilian Jiu-Jitsu, founding a Brazilian Jiu-Jitsu school along the way.

Waitzkin did not stop with being a chess prodigy and a martial arts champion. He is also a successful author, having authored two books, "Attacking Chess: Aggressive Strategies, Inside Moves from the U.S. Junior Chess Champion" in 1995, and "The Art of Learning: An Inner Journey to Optimal Performance" in 2008.

Josh Waitzkin and the remarkable diversity of his accomplishments in three very distinct fields that required exercising of three very different skills is a testament that any individual can learn any new skill, but has to detach from the learnings of the previous one. There is little that Josh Waitzkin could have applied from chess to martial arts. There is little that he could carry forward to being an author from either chess or martial arts. Each of these three fields are vastly different from each other. To become adept at each new field – let alone achieve the upper echelons of that field – not only requires incredible focus and learning skills in that new field, but also requires an individual to disconnect and detach from all they have known until that point in time. Their comfort zone. It is without doubt that stepping into a diametrically distinct field such as martial arts required Waitzkin to step outside his comfort zone – chess up until that point in time – and pivot towards learning something brand new, decoupled from all that he had learned until then. With little to carry forward from his previous field, and to become a World Champion within an entirely new field in a compressed span of time, Waitzkin had to unlearn all that he had learned, such that he could learn something brand new - no matter how dramatically different this brand new thing that he learned was. Those who can unlearn existing habits and "muscle memories" in order to accommodate making room for learning something new, can demonstrably be successful at it, in a relatively rapid pace.

3. Relearn – Desmond "Lightning" Douglas MBE

Sir Desmond Douglas, nicknamed "Lightning", is widely considered to be Great Britain's most successful table tennis (ping pong) player in the last fifty years. An 11-time English champion, Douglas was renowned for his lightning fast reflexes. With an unorthodox style of taking the ball early, his playing style was just as unorthodox. Douglas played particularly close to the ping pong table and looked self-restricting in his mobility and his ability to leverage the full extent of the space and distance around him.

Competing beginning at the young age of 15, Sir Douglas had risen to the top of the British National Junior Rankings by the age of 18. At 18, he also won his first European Junior title. In an astonishing feat, Desmond Douglas conducted a clean sweep at the English National Championships in 1976, winning singles, doubles, and mixed doubles titles. He followed up by winning English, and Welsh Open titles, and winning bronze at the European Championships in 1978. It was at this time that he rose to be seeded at number 7 in world rankings. Winning several additional titles in Germany, Sir Douglas returned to England in 1985. He was honored with the MBE in 1987 for his significant contributions to his sport. The MBE, short of Member of the Order of the British Empire, is the third highest ranking Order of the British Empire award, and bestows the honorific "Sir" title upon its recipient. In 1988, Douglas competed in the Seoul Olympics, competing in singles competition and earning a ninth-place finish in the doubles category. After retiring briefly from the sport in 1989, Douglas came out of retirement and competed into his forties.

The 7th best in the world and 11 time English champion, renowned for the speed at which he played, drew the attention of researchers from Brighton University who wanted to study his lighting fast reflexes. The researchers staged a reactions test to measure Douglas' reflexes speed, a test that included and also measured response and reflex times of other members of the English Table Tennis team, including players who were several years younger than Douglas. To the astonishment of the researchers,

Sir Douglas' reflex times were actually the worst amongst the entire set of English Table Tennis athletes that were tested by the University. As a direct consequence of this result, the test itself faced derision, and was written off as flawed, the premise being that no one could believe that someone who played at such lightning speeds could never have the worst reaction times of any of his peers.

However, as it turned out, the test was not a failed experiment. Douglas did indeed have the slowest reaction times amongst his English Table Tennis peer group, despite playing significantly faster than all of them. Intrigued by this paradox, the University study delved deeper into how it was possible for Douglas to be lightning fast and yet have the slowest reflexes of any players on the English Table Tennis team. An investigation into his background and how he carried that forward into his approach to the game provided the answers.

Sir Douglas grew up playing the sport in a cramped room, with very little room available to him behind the table. It was due to this reason that he had to play so close to the table with his stomach pressed up against the table. These cramped and confined playing quarters forced him to respond quickly and hit the ball early. Without this approach, Sir Douglas would have not been able to return a serve. While this might seem to be a disadvantage to most, the confinement actually served him as an advantage to how he approached the game. Having become adept at playing the way he did – with hitting the ball early – when he carried his learnings to a normal table, he exhibited what seemed like lighting fast reflexes.

With each serve lobbed at him, to hit the ball early, it was imperative that Sir Douglas was able to accurately predict where the ball would land. As any tennis, squash, or table tennis player knows, this is an incredibly difficult task in the midst of active gameplay. Being able to predict where the ball is going to land, such that he could reach the ball early, required Sir Douglas to unlearn the immediately preceding rallies and relearn how to adjust his stylistically limited physical span to connect with the ball. It wasn't that Sir Douglas had the fastest reflexes on his team, it was that Sir Douglas was much more adept at learning his opponent's style, unlearning the immediate preceding move, and relearning where the next serve was going to land.

A Lifelong Learning Mindset vs. A Learn, Unlearn, Relearn Mindset

The next several decades are going to require us to embrace lifelong learning, and a continual learning mindset will be crucial for tomorrow's job market. Continual learning - being a lifelong learner - is table stakes in this digital age. In a rapidly evolving job market, learning has to be a lifelong pursuit. This continual learning mindset, also referred to as a growth mindset is essential in an era of rapid change. By developing a growth mindset and embracing a commitment to ongoing education, we can stay ahead of technological advancements and equip ourselves with the skills demanded by future roles. This approach ensures that our skills remain relevant and enables us to adapt to emerging trends and technologies. Continuous learning will be crucial in navigating the ever-evolving landscape of AI and technological change. Adaptability is crucial in the face of career risks posed by AI and technological change. It will be imperative to prioritize lifelong learning, and embracing opportunities to acquire new knowledge, skills, and competencies. By actively engaging in learning platforms, participating in educational programs, and seeking diverse experiences, each of us can remain agile, adapt to emerging trends, and capitalize on new opportunities.

We have to embrace the belief that skills can be developed – like the Polgar sisters and Josh Waitzkin demonstrated – anyone can learn and become really good at anything. We have to embrace change as an opportunity for growth and actively seek out opportunities to diversify skill sets. Not everything will be perfect along this journey, and we need to recognize that we will make missteps and mistakes. There is no playbook for the era we are heading into. And that is ok; failures are ok. We must invest in resetting our thinking around failure, and seeking failures as opportunities for learning and for growth. We have to cultivate resilience to bounce back from setbacks and persevere in the face of uncertainty. A 21st century growth mindset will require us to view AI and technological advancements as catalysts for personal and professional development, rather than as insurmountable obstacles.

All of these things that we can and should do to embrace continual learning like we haven't before are critical. Continual learning is at the heart of being able to adopt and adapt. However, in and of itself, continual learning is <u>**inadequate**</u> to "**thurvive**" the Fourth Industrial Revolution.

In my view, I see the continual learning mindset and the "learn, unlearn, and relearn" mindset as interrelated, yet quite distinct, differing in their approach and focus. Both mindsets have their place and relevance in the age of digital and technology. The continual learning mindset provides a solid foundation for ongoing personal and professional growth, enabling us to stay informed and develop expertise in their fields. Meanwhile, the "learn, unlearn, and relearn" mindset adds an additional layer of adaptability and agility, equipping us to navigate and "**thurvive**" amidst rapid change and uncertainty.

The continual learning mindset revolves around the concept of lifelong learning. It emphasizes the importance of consistently acquiring new knowledge, skills, and perspectives throughout one's career. Those with a continual learning mindset understand that learning is a continuous journey, not limited to formal education or specific stages of their career. They actively seek out opportunities for growth – via books, courses, workshops, or experiences. This mindset is characterized by curiosity, adaptability, and a hunger for knowledge, as individuals strive to expand their intellectual horizons and stay ahead in a rapidly evolving world.

On the other hand, the "learn, unlearn, and relearn" mindset recognizes that in the face of constant change, simply accumulating knowledge is just not enough. It involves a more dynamic, proactive, and transformative approach. While learning is still valued, this mindset acknowledges the necessity of ***unlearning*** outdated or ineffective beliefs, skills, and habits. It requires individuals to critically examine their existing knowledge and practices, challenge assumptions, and be open to letting go of what no longer serves them. ***Relearning*** then comes into play, as individuals actively seek new information, update their skills, and adapt to emerging trends and technologies. The focus here is on agility, flexibility, and the ability to adapt to shifting contexts. The "learn, unlearn, and relearn" mindset encourages individuals to embrace change, adapt quickly, and continually update their approaches to remain relevant in dynamic environments.

In essence, the key distinction lies in the level of adaptability and transformation involved. The continual learning mindset emphasizes a continuous accumulation of knowledge and skills, building upon existing foundations. It is more evolutionary in nature, allowing individuals to stay relevant and grow incrementally. On the other hand, the "learn, unlearn, and relearn" mindset embraces a more revolutionary approach. It requires us to actively challenge and reshape our mental models, embracing change and transformation as an integral part of our learning journey. While the continual learning mindset emphasizes the importance of ongoing knowledge acquisition and growth, the "learn, unlearn, and relearn" mindset takes it a step further by encouraging us to challenge existing beliefs, ***unlearn*** obsolete knowledge, and actively embrace change. The "**learn, unlearn, and relearn**" mindset is particularly relevant in Fourth Industrial Age, where agility, nimbleness, innovation and disruption are the norm.

The Thurvival Mindset

In this journey through the transformative power of technology, and the importance of "**learn, unlearn, and relearn**" mindset, we have explored the profound impact of the upcoming technological revolution on our lives, and the imperative of evolving our mindset from one focused purely on lifelong learning. Each of us harbors limitless potential within us to adopt and adapt, innovate, and leverage technology for personal and professional growth. In the age of digital and technology, the need to rapidly learn, unlearn, and relearn has become more crucial than ever before. The rapid pace at which information and technology evolve necessitates a mindset of continuous learning, unlearning, and relearning.

This "**learn, unlearn, and relearn**" mindset will is a fundamental requirement for our professional growth in the Fourth Industrial Revolution. The significance of embracing continuous learning, challenging outdated beliefs, and remaining adaptable in the face of technological advancements cannot be overstated to navigate the ever-changing landscape with confidence, seize new opportunities, and contribute meaningfully to a digitally transformed world. In this age of abundant information, it is not enough to simply acquire new knowledge. We must also be willing to let go of outdated or inaccurate beliefs, ideas, and practices. Unlearning involves challenging our preconceived notions, questioning long-held assumptions, and being open to alternative perspectives. By shedding obsolete knowledge and mental frameworks, we create room for new ideas and innovative thinking to flourish. However, the process of unlearning alone is not sufficient. Relearning complements the unlearning process by enabling us to acquire updated and relevant knowledge. Relearning involves seeking out new information, developing new skills, and adapting to emerging paradigms. It requires a mindset of flexibility and adaptability, as well as a willingness to embrace change and explore unfamiliar territories. Relearning allows us to integrate new knowledge into our existing skill sets and adapt our strategies accordingly, thus ensuring our continued growth and success in the digital age.

By embracing a mindset of continuous learning, challenging our assumptions through unlearning, and adapting to change through relearning, we can navigate the ever-evolving digital landscape with confidence and thrive in an age of endless possibilities. This mindset is the key to unlocking our full potential in the digital era.

Chapter Seventeen – Thurvival Technique 2: Harness Technology

There is a proverb that says "if you can't beat 'em, join 'em". Unlike the luddites of the past, we don't have to beat technology in order to "**thurvive**", nor should we harbor any expectations that this foolhardy quest would have a modicum of success. Rather than thinking about stopping or even slowing down the accelerated march of technology's progress as pertains to our careers, we should be thinking about how to capitalize on it. Technology can be our greatest asset as a vibrant resource, or our largest liability as a source of disruption and disintermediation. Those who can successfully take advantage of technology will realize opportunities to thrive in the Fourth Industrial Revolution.

This book explored the remarkable progress we have made in just the last century alone, the entirety of this progress attributable to the quantum leaps of technological advancements. This unprecedented explosion of technological progress is responsible in transforming the way humans live, work, and interact with the world. Earlier in this book, we touched upon the modern conveniences that were unimaginable just a hundred years ago. From the way this very book is authored and published, from communication to transportation, healthcare to entertainment - the impact of technology has revolutionized every aspect of human existence.

As we think about the next twenty-five years of technological progress and the impact to our professions, let's revisit some of the notable modern conveniences that were developed in the past century. A hundred years ago, long-distance communication was limited to telegrams and postal mail. Today, we enjoy the convenience of instant global communication at our fingertips. Mobile phones, the internet, and social media platforms have connected us in ways that would be unimaginable to people a hundred years ago. We can video call loved ones across continents, collaborate with colleagues in real-time, and share experiences instantly through social networks. The world has become a global village, fostering connections and cultural exchange on an unprecedented scale.

Access to knowledge was confined to libraries and limited educational institutions a hundred years ago. Today, the internet has transformed information access beyond imagination. Search engines provide instant answers to our questions, and online encyclopedias serve as vast repositories of knowledge. Generative AI platforms such as ChatGPT are poised to further revolutionize how we secure access to knowledge and information. Online courses and e-learning platforms offer educational opportunities to millions, democratizing access to education. The ability to access and share information freely has propelled humanity's collective knowledge and accelerated progress across various fields.

Advancements in medical technology over the past century have revolutionized healthcare and extended human lifespans. Life expectancy at birth at the turn of the 20th century, was 46 years for men and 48 years for women. In 1776, at the time of America's founding, the average American citizen could expect to live to the age of 35. Contrast that to the early 2020s - the average life expectancy in the United States is around 74 years for men and 80 years for women. Diagnostic tools such as MRI scans, CT scans, and genetic testing have enabled early disease detection and personalized treatments. Minimally invasive surgeries, robotic-assisted procedures, and remote healthcare monitoring have improved patient outcomes and reduced recovery times. Telemedicine and remote patient monitoring have brought healthcare services to underserved areas and enabled virtual consultations, improving access and convenience for patients. Wearable devices and health tracking applications empower individuals to proactively monitor their well-being, and have allowed in the promotion of preventive care. Breakthroughs in pharmaceuticals, vaccines, and gene therapies have eradicated diseases and enhanced overall well-being, contributing to longer, healthier lives. The COVID-19 vaccine was developed in record time for any vaccine, largely due to the significant improvements in technology – a feat unimaginable in 1918, during the last global pandemic. Note that it would be another decade after the last global pandemic for the discovery of antibiotics, with Alexander Fleming's discovery of penicillin in 1928 in London. This discovery led to the introduction of antibiotics that greatly reduced the number of deaths from infections.

We discussed earlier in this book how remarkable it is that it took only 66 years from Wright Brothers taking flight to humanity landing on the moon. Consider how much of a metamorphosis transportable has gone through, effectively transforming the way we navigate the world. A century ago, long-distance travel was limited to slow steamships and trains. Today, air travel allows us to traverse the globe in a matter of hours, with high-speed trains, electric vehicles, and Hyperloop technologies offering efficient, eco-friendly modes of transportation. Ride-sharing platforms and navigation apps have made commuting more convenient and streamlined, negating the need to own an automobile in cities. The transformation of transportation has allowed humans to explore the world with ease, connecting cultures and broadening our collective horizons.

The concept of a smart home, where appliances and devices can be controlled remotely, was unimaginable a hundred years ago. Today, we have voice-activated virtual assistants, smart thermostats, connected security systems, and automated household appliances. With a simple command, we can adjust the lighting, control the temperature, and even order groceries. Home automation has made our lives more convenient, energy-efficient, and secure, transforming our living spaces into intuitive, personalized environments. Virtual assistants like Amazon's Alexa and Google Assistant have become personal concierges, simplifying our daily routines and enhancing convenience. Automated thermostats, energy-efficient appliances, and home security systems have not only improved our quality of life but also helped reduce energy consumption and promote sustainability.

The world of entertainment and media has undergone a seismic shift. A century ago, radio and silent movies were the height of technology. Today, we have high-definition television, streaming services, virtual reality, and augmented reality experiences. Online platforms provide instant access to an endless array of movies, music, and literature. Social media and content-sharing platforms have given rise to a new era of user-generated content and digital influencers. Entertainment has become immersive, personalized, and accessible to all. Gaming has also undergone a transformative evolution, with immersive virtual reality experiences and realistic graphics captivating players in breathtaking digital worlds. These advancements have made entertainment more personalized, convenient, and engaging than ever before.

Benefits of Technology as a Career Ally

The progress of technology has transformed our lives in unimaginable ways, ushering in a world of conveniences that would have seemed like pure fantasy a mere century ago. How often do you talk with your coworkers about wanting to go back in time and buy that Google stock or invest in Tesla? Now consider what your life as a professional would have been – what your odds of career success could have been – had you been able to ride the crest of the technology wave in relation to any single one of the aforementioned innovations over the past century? Some – such as Larry Page and Sergei Brin of Alphabet/Google, Mark Zuckerberg of Facebook/Meta, and Elon Musk of Tesla/SpaceX/Twitter – did not have to imagine. They were on the leading edge of leveraging technological progress to bring about societal change. While your own ambitions might be more conservative than any of these individuals, there are just as many opportunities awaiting you at whatever scale you choose.

The rapidly evolving landscape of technology has given rise to a new paradigm in which humans can harness its power to augment their jobs and advance their careers. Whether you want to be a technology entrepreneur, ensure you're keeping your skills current and fresh, or simply ensure you're going to future-proof your career, it is vital to identify ways you can employ and deploy technology. My friend's mother, whose story we visited in Chapter Fifteen, is a testament to a successful professional who found success by continually leveraging technological advancements to her career advantage.

Harnessing technology as an ally enables us to unlock new levels of productivity, efficiency, and innovation. Thurvival Technique 2 – Harness Technology - explores the compelling reasons why we should leverage technology in a professional environment to augment our jobs and propel our careers to new heights. The integration of technology into professional environments is no longer a choice but a necessity for career advancement. By harnessing the power of technology, we can amplify our productivity, expand our skill sets, embrace remote work, engage in continual learning, unleash our creativity, enhance decision-making, and navigate the digital transformations abound across our companies. The future belongs to those who embrace technology as a tool to augment their jobs, better their careers, and remain agile in an ever-changing world. It is a bedrock for "**thurvival**". Not that you needed further convincing that technology will fundamentally improve your own job today and your career over the long-term. There is significant value to be created by capitalizing on technological advancements within your own job.

Technology offers a myriad of tools and applications that can significantly enhance productivity in the workplace. From project management software that streamlines workflows to automation tools that handle repetitive tasks, we can offload mundane and time-consuming activities, allowing us to focus on high-value, strategic work. By leveraging technology, we can accomplish more in less time, increasing our output and driving our careers forward. Automation has been a cause for concern, but when embraced intelligently, it can liberate us from mundane, repetitive tasks, allowing them to focus on more strategic and creative endeavors. Robotic Process Automation (RPA) can automate routine data entry, report generation, and administrative tasks, freeing up valuable time for us to engage in critical thinking, problem-solving, and innovation. By leveraging automation, we can streamline workflows, reduce errors, and increase overall productivity. Mundane and repetitive tasks can be time-consuming and prone to errors. AI-powered automation can handle such tasks, freeing us up to focus on higher-level responsibilities that require critical thinking, creativity, and problem-solving. This enables us to work more efficiently and dedicate our time to tasks that truly require our unique cognitive abilities.

In a world driven by technology, acquiring and honing technical skills is essential for professional growth. Embracing technology in the workplace creates opportunities for us to expand our skill sets and stay relevant to a step ahead in our respective fields, similar to my friend's mother, the technologically progressive seamstress. Whether it's learning programming languages, mastering data analysis tools, or familiarizing ourselves with emerging technologies like AI, integrating technology into our jobs will open doors to new avenues of professional development. In an era of rapid technological advancements, continual learning is crucial for career advancement (as part of the **Learn-Unlearn-Relearn** mindset). Technology provides an abundance of resources and platforms to engage in lifelong learning. Online courses, webinars, and virtual training programs offer flexible and accessible avenues for acquiring new knowledge and skills. E-learning platforms, Massive Open Online Courses (MOOCs), and virtual classrooms offer flexible, accessible opportunities for acquiring new skills, expanding knowledge, and staying abreast of industry trends. Personalized learning algorithms can tailor educational content to individual needs, maximizing learning outcomes. By leveraging technology as a learning tool, we can continuously upskill, adapt to changing work dynamics, and remain competitive in the job market. By embracing technology as a learning tool, we can stay ahead of the curve, adapt to changing industry trends, and remain competitive in our careers.

As proven by the COVID-19 pandemic and remote work on a massive scale, technology has revolutionized the way we work, enabling remote collaboration and flexible work arrangements. We can leverage communication and collaboration tools to work with teams across different time zones and locations, breaking down geographical barriers. The ability to work remotely and have flexible schedules has not only enhanced work-life balance but also allowed individuals to explore new opportunities and pursue career growth without being confined to a specific location. Effective collaboration and communication are fundamental for success in any professional setting. Technology offers an array of tools and platforms that facilitate seamless interactions, regardless of physical boundaries. Cloud-based collaboration tools enable real-time document sharing, editing, and commenting, fostering efficient teamwork. Video conferencing and instant messaging platforms bring together geographically dispersed teams, enabling face-to-face interactions and fostering a sense of connection. Additionally, social media platforms and professional networking sites provide avenues for knowledge sharing, idea generation, and peer-to-peer learning. The past two decades have revolutionized the way that we connect, network, and seek career opportunities. Online platforms, social media networks, and professional networking sites provide avenues for establishing valuable connections with peers, mentors, and industry leaders from around the globe. These digital networks open doors to collaborations, partnerships, and career advancements that would have been otherwise inaccessible. By harnessing technology for networking purposes, we can expand our reach, tap into a diverse talent pool, and explore new horizons of career growth.

Technology serves as a powerful catalyst for creativity and innovation. Advanced software, virtual reality, and augmented reality tools can empower us to explore new ideas, visualize concepts, and prototype designs. Digital platforms offer avenues for collaboration and idea sharing, fostering a vibrant ecosystem where innovation flourishes. By harnessing technology, we can push the boundaries of our creativity, find novel solutions to complex problems, and differentiate ourselves within our fields. Digital tools and software can empower us to explore new ideas, visualize concepts, and prototype designs. Advanced 3D modeling and simulation tools enable engineers, architects, and designers to bring visions to life virtually, accelerating the product development lifecycle. Artificial intelligence (AI)-driven algorithms can generate novel solutions, inspire artistic creations, and support content generation, supplementing human creativity with machine intelligence.

In any industry, the ability to gather, analyze, and interpret data is crucial. By way of data and analytics, technology has revolutionized the way we access and manage information. By utilizing advanced data analytics tools, we can quickly derive meaningful insights, make informed decisions, and identify emerging trends. Intelligent algorithms and machine learning systems can process vast amounts of data, providing us with a comprehensive understanding of complex situations. Augmented Reality (AR) and Virtual Reality (VR) technologies further enhance information visualization, enabling us to immerse ourselves in virtual environments, collaborate remotely, and gain new perspectives. In the age of big data, technology equips us with the tools to make data-driven decisions. Advanced analytics and visualization tools enable us to gather, analyze, and interpret vast amounts of information, leading to more informed and strategic decision-making. By analyzing large datasets and identifying patterns, AI algorithms can offer data-driven recommendations, helping professionals in various fields, such as finance, healthcare, and marketing, to make more accurate and efficient decisions.

By leveraging technology, we can derive meaningful insights, identify trends, and anticipate market shifts, ultimately gaining a competitive edge and advancing our careers. By integrating AI-powered analytics and predictive models into decision-making processes, individuals can make more informed choices based on data-driven evidence. This collaboration allows humans to leverage the analytical capabilities of AI while bringing their unique contextual understanding, intuition, and ethical considerations to the table. AI technologies, such as natural language processing and machine learning algorithms, can analyze vast amounts of customer data to personalize interactions and provide tailored recommendations. This enables professionals in sales, customer service, and marketing to deliver more personalized experiences, fostering stronger customer relationships and driving business growth. Let's go back to where this book started – where your spouse, an insurance agent, made a windfall because her AI algorithms were able to predict, with a level of accuracy, which customers would most benefit from what type of an insurance product. These predictions are predicated on big data and AI.

Augmented Intelligence is the other "AI" that we don't think about. Specifically, how do we think of technology, and the deployment of technology to augment human intelligence in order to better our jobs? It's going to be tempting for us to fret about AI – Artificial Intelligence – disrupting our professions. Thurvival Technique 2 implores us to consider the other AI, "Augmented Intelligence" – by augmenting ours own job by the successful application of technology. As the rapid advancement of artificial intelligence (AI) and technology reshapes industries and job markets, there is a growing realization that we can collaborate with technology to achieve unprecedented levels of productivity, efficiency, and innovation. Augmenting our jobs with technology empowers us to leverage the strengths of both humans and machines, creating a synergy that drives progress.

Human-Technology collaboration is not without its challenges. Trust-building between humans and machines is essential for effective collaboration. Transparent and explainable AI algorithms, ethical considerations, and clear communication about the roles and limitations of AI systems are crucial in fostering trust. It is imperative to address concerns around job displacement and ensure that AI augmentation is implemented with a human-centric approach, focusing on empowering individuals rather than replacing them. Beyond augmentation, we can actively participate in the co-creation of AI systems, contributing our domain expertise, contextual knowledge, and ethical considerations. Collaborative efforts between us and technology can lead to the development of more inclusive, fair, and unbiased AI systems that align with our values and societal needs. This collaborative approach ensures that we remain in control of the technology we create, shaping it to serve our best interests. Industries across the board are undergoing a digital transformation. Those who embrace technology and develop digital literacy have a significant advantage in adapting to these changes. By harnessing technology, we can navigate the evolving corporate landscape, leverage emerging opportunities, and future-proof our careers. We can become agents of change within our organizations, leading digital initiatives, and driving innovation.

Harnessing Technology

Harnessing technology offers immense potential for us to augment our jobs, increase productivity, and drive innovation. Rather than viewing technology as a threat, we should embrace it as a powerful tool that complements and amplifies our professional capabilities. Technology excels at processing vast amounts of data, identifying patterns, and automating repetitive tasks, while we possess unique qualities such as creativity, intuition, emotional intelligence, and complex decision-making abilities. By combining the strengths of both – us, as workers – and technology, we can achieve unprecedented levels of productivity and problem-solving.

There are a plethora of ways that technology can be deployed TODAY within a myriad of professions. If you are employed in any of these industries, there are an unlimited number of opportunities available to you to immediately employ technology within your own job. In the field of healthcare for instance, AI can be deployed to assist medical professionals in diagnosing diseases, analyzing medical images, and predicting treatment outcomes. By leveraging technology such as AI, physicians can access valuable insights and recommendations, leading to improved patient care and more accurate diagnoses. Chatbots also enhance patient support and provide personalized health information. In manufacturing today, in addition to the automation of the assembly line, technology can enable smart factories and manufacturing processes. Collaborative robots, known as cobots, work alongside humans to increase efficiency and safety. AI algorithms can analyze production data, optimize supply chains, and predict maintenance needs, leading to reduced costs and improved productivity.

In Finance, technology can assist with fraud detection, risk assessment, and investment strategies. AI chatbots and virtual assistants can handle customer inquiries, while AI-powered algorithms analyze market trends and patterns for more informed decision-making. In the field of Education, technologies such as AI tools can personalize learning experiences, providing tailored recommendations and adaptive learning paths. Intelligent tutoring systems can offer personalized feedback, while automated grading systems streamline the assessment process. Educational platforms can analyze data to identify areas of improvement and optimize learning outcomes. Creative fields such as the arts, music, theater, etc. are considered to be intrinsically human. While this is true, it does not imply that professionals in those fields are without a considerable number of opportunities for leveraging technology within their companies. AI, including Generative AI such as ChatGPT and DALL-E can augment creative fields such as art, music, and design. Machine learning algorithms can generate new artistic styles, assist in music composition, and help designers explore innovative concepts. By collaborating with AI tools, human creatives can push the boundaries of their own imagination and produce groundbreaking works.

Regardless of the field and industry that you are employed in, there are several practical and pragmatic ways that we can harness technology and incorporate technological advancements in our careers:

1. Staying Informed:

Human evolution is marked by the ability of our ancestors to leverage tools. Although Generation Z, the first digitally native generation, demonstrates an uncanny ability to be proficient with technology (particularly consumer electronics) without training, the ability to harness technology can be a learned skill, as opposed to a natural ability. Just as with the "**learn, unlearn, relearn**" mindset, we must embrace a growth mindset when it comes to technology, staying abreast and informed of technology advancements, understand its potential applications, and explore relevant training opportunities – especially in our field or industry. By continuously updating our skills and knowledge, we can remain adaptable and leverage technology effectively in our respective industries.

2. Identify Opportunities:

Identify tasks or processes within your job that can benefit from technology or AI augmentation. Look for repetitive, data-intensive, or time-consuming tasks that can be automated or enhanced through a myriad of technologies. Leverage tools and platforms to automate routine tasks, extract insights from data, and generate recommendations. Embrace technology as a partner that complements your skills and amplifies your capabilities. Imagine making yourself the go-to resource within your organization for all things technology – someone who has been building a reputation of harnessing technology for making the operation efficient, automating, reducing errors, increasing profitability, attracting and retaining new customers by using analytics, etc. That is a sure way to stand out within your own firm and developing a robust career for yourself, regardless of how rapidly technology progresses around us.

3. Collaborative vs. Cautious and Contentious:

It goes without saying – we simply cannot afford to become the Luddites of the 21st century. We have to embrace a collaborative mindset when it comes to technology in the workplace, and actively seek opportunities to collaborate with workplace systems. Instead of taking a passive or detractor approach to new technology introduced at work, we have to lean in and engage with these technologies, understand their capabilities, and identify areas where they can enhance our work. We should gain knowledge and understanding of technologies relevant to your field, staying updated on the latest advancements and explore how they can be applied to our work. Online courses, workshops, and industry conferences can provide valuable insights and resources. Building a collaborative partnership between us and technology will ensure the best outcomes for our careers. We cannot afford to squander time and opportunities in being overly cautious, or worse yet, demonstrate contentiousness and resist the introduction and implementation of new tools and technologies within our workplaces.

4. Be a Domain Expert:

No matter how much technology progresses, our deep domain knowledge remains vital to the success of our organizations. While technology can automates, process data, and generate insights, it is our contextual understanding and expertise that enable effective interpretation and application of technology, specifically AI recommendations. Developing and maintaining expertise in our field of work will be critical, as it allows us to guide and optimize the contributions that technology makes most effectively.

5. Ethics:

We should familiarize ourselves with the ethical implications of technology, especially with things such as AI. We should stay informed about ethical guidelines and best practices, and actively advocate for transparency, fairness, and privacy. Building your awareness of ethical considerations will leverage your HUMAN expertise and judgment, and ensure that technology is deployed responsibly and for the benefit of society.

6. Adopt and Adapt:

As you work alongside technology, continuously assess and refine how the technology is performing within your enterprise. Provide feedback, fine-tune the technology implementation, and adapt the platforms to better suit your specific needs. The synergy between human expertise and technological capabilities should be a continuous cycle of improvement.

7. Emphasize YOUR Human Skills:

As technology can handle repetitive tasks, focus on developing and emphasizing uniquely human skills such as critical thinking, creativity, emotional intelligence, and complex problem-solving. These skills will be increasingly valuable as technology augments our work, enabling us to provide unique insights and make complex decisions. This is the crux of Thurvival Technique number 4, and shall be explored in detail in Chapter Nineteen.

Drive your Digital Destiny

The ability to embrace and integrate emerging technologies into one's skill set is an essential trait for career success in the digital era. The integration of technology into professional environments is no longer a luxury but a necessity for career advancement. In an ever-evolving world driven by technological advancements, the role of us - humans - in the professional landscape is constantly transforming. While the fear of automation and job displacement due to technological advancements has been a recurring theme throughout history, the future holds immense potential for us to leverage technology as a powerful tool for augmenting our jobs. By embracing and harnessing technology, we can enhance our productivity, creativity, and efficiency, ultimately leading to improved job satisfaction and higher quality outcomes. By embracing technology as an ally and as a collaborative tool, we can create new opportunities, explore uncharted territories, and navigate the ever-evolving landscape of work with confidence and adaptability. The future lies in the synergy between us and technology, where our collective strengths combine to create a world of endless possibilities. Technology has brought about significant transformations across various industries, and it's only just getting started. The integration of technology into our professions presents an opportunity for us, as professionals, to augment our jobs, unleashing our full potential by harnessing the power of technology. By leveraging technological advancements, we can revolutionize our workplaces and achieve new levels of productivity.

Through responsible augmentation, we can carve out a future for ourselves where technology enables us to focus on more complex and strategic aspects of our work, and continues to act as a catalyst for innovation and in turn, nurtures an environment where our human ingenuity flourishes. This symbiotic relationship between humans and technology holds the key to a future where professionals are empowered to adapt, learn, collaborate, and thrive in an ever-evolving digital landscape. The possibilities are vast, and the time to embrace technology as a professional ally is now. By harnessing technology, we can enhance our skills, unlock new possibilities, and "**thurvive**" in the digital age by being in the driver's seat of our own digital destinies. Augmenting your own career with technology, keeping abreast with the latest technological developments, and ideating and experimenting with these tools within your workplace is vital. To ensure that you are directing and channeling your technological learnings appropriately, you need to be a "student of the sport". You need to be fluent in the language of business. In other words, knowing the technological direction of your profession can be exponentially strengthened by Thurvival Technique 3. Whereas, by learning and adopting technology to your job, you might become the most technologically adept employee, you don't want to be the most technologically adept employee at Blockbuster in the year 2010. As Chapter Eighteen details, it is essential for you to know the direction that your industry and company is taking.

Chapter Eighteen – Thurvival Technique 3: Understand Your Industry And Company Trends

Companies and entire industries often find themselves at a proverbial fork in the road during these change-induced revolutionary periods. I label the behavior of companies driving, acting, or reacting to the changes in one of six ways. Consider these six categories to be on a linear spectrum, where the lower the number (1 through 6), the better off a company's chances of driving the change or thriving amidst change. Bookending either end of this spectrum - on one end are organizations that have the vision to drive the change or the innovation within an entire sector, others with the foresight to find incredible opportunities within these emerging advancements and become juggernauts by seizing on these advancements - while on the other end of this spectrum are others shy that away from them. With due apologies to my elementary school grammar teachers (and all the engineers at Microsoft who toiled on the Spell Check feature in Microsoft Word), incorrect spelling of some of these labels is intentional.

Note that when I refer to "macro changes" here, I am alluding to broader macroscopic changes to industries and companies that arise as a result of an industrial revolution. When I refer to "business changes", I am speaking to the changes in direction, updates to business plans and strategies, or recalibration of corporate objectives at a company level; these business changes as a direct result of macro changes.

Forecasting The Future, Predicated On The Past

i. Visionaries

Visionary companies are helmed by an individual or a group of individuals that invent, introduce, or implement a tool, technology, or technique that transforms the direction of an industry – and, without hyperbole, at times of human history. They serve as the drivers of change and the protagonist/s of an Industrial Revolution. These are individuals who create and implement something new and innovative and entire industries or even economies are built on their having served as the catalyst to change.

Few notable examples include the discovery and subsequent application of electricity. Benjamin Franklin is given the credit for discovering electricity. In 1752, Benjamin Franklin conducted an experiment using a kite and key on a rainy day to demonstrate the relationship between lightning and electricity. Having flown a kite tied with a key during the thunderstorm, the electricity from the storm clouds transferred to the key and he received an electric shock. The first practical application of electricity was the invention of electric dynamo by Michael Faraday in the 1831. Faraday's invention served to trigger an electrical revolution around the world. Nikola Tesla invented alternating current and induction motor in the late 1800s.

According to a report that appeared in Springer Science and Business Media in 2000, "After devising a commercially viable electric light bulb on October 21, 1879, Edison developed an electric "utility" to compete with the existing gas light utilities. On December 17, 1880, he founded the Edison Illuminating Company, and during the 1880s, he patented a system for electricity distribution." (Faruqui & Eakin, 2000). Elon Musk (Founder and Chief Executive Officer of Tesla Motors and SpaceX) founded Tesla Motors to compete with major automobile manufacturers. Combustion engine vehicles, running on gasoline had been the norm for over a hundred years since Ford's Model-T production in the 1908.

Telsa Motors disrupted the fossil fuel operated automobile industry with their cutting edge, technologically advanced fleet of vehicles. As of 2021, most major automobile manufacturers boast a robust upcoming fleet of all-electric vehicles, while some have declared a specific date when they will cease manufacturing combustion engine vehicles in favor of a fully-electric fleet. Had it not been for Telsa Motors leading the way (note that there were other electric automobile manufacturers that came before Tesla, but none took hold the way Telsa did), proving the commercial viability of all-electric vehicles, this sea change would have likely been years away.

ii. Capitalizers

Capitalizers are companies that are helmed by visionaries who, inspired by the vision of the visionaries, take a risk in driving their firms to adoption of the new technology, tool, or technique. While they might not necessarily be the inventors of these innovations, they have the ability to recognize the transformation potential that these inventions hold, what it means for the industry they occupy and ultimately what it means for their firm. They are often seen as early adopters (retrospectively), but are often subject to skepticism – or worse – scorn and ridicule – at the particular point in time. They see what others do not see. They take the risks that others do not take. They are equitably responsible for transforming an industry as much as the Visionaries. A good vision, absent someone to buy into it and propagate it, can prove limiting.

There are several examples of Capitalizers through history. In combination with Visionaries, they form a potent source for the widespread distribution and adoption of the change. Since we explored the genesis of generating and distributing electricity and the Visionaries who made this possible, staying with the same theme is an example of American businesswoman and a pioneer in her field, Marjorie (nee: Post) Merriweather. While Ms. Merriweather, who died in 1973 at the age of 86, is most well-known for being the owner of General Foods, the erstwhile

Postum Cereal Company (better known as Post Cereal). Ms. Merriweather, along with her second husband, E.F. Hutton, began expanding Postum Cereal Company, acquiring other American food companies such as Hellmann's Mayonnaise, Jell-O, Baker's Chocolate, Maxwell House (NYT, Mrs. Marjorie Merriweather Post Is Dead at 86, 1973), and many more before the conglomerate was renamed as General Foods in 1929. One of her acquisitions, a company called General Seafood Corporation, helmed by inventor Clarence Birdseye that had developed an innovative way to quick freeze food products earned her a fair share of skeptics at the time. A leading skeptic of her interest in this purchase was her husband E.F. Hutton, who did not see the value of frozen food.

In retrospect, Merriweather's purchase of Birdseye's company can be recognized for her pragmatic risk-taking, recognizing the sheer untapped potential of a new revolutionary development in food processing. Clarence Birdseye was an American inventor, entrepreneur, and naturalist, considered the founder of the modern frozen food industry (Wikipedia, Wikipedia, n.d.). In 1924, Mr. Birdseye developed a new process for commercially viable quick-freezing; packing fish in cartons, then freezing the contents between two refrigerated surfaces under pressure. However, since electricity itself was available in less than half of US homes in 1924, refrigeration, and specifically, freezers, were not available to most of the population. Additionally, in addition to frozen food being seen as some kind of a novelty item, there was the issue of transport, distribution, and storage, in addition to most consumers not having access to be able to purchase and utilize these products.

This lack of consumer interest almost caused Mr. Birdseye to go bankrupt. Birdseye kept inventing and innovation on his concept. In 1925, he invented the double belt freezer, an invention that was issued a US Patent and is generally considered as the beginning of today's frozen foods industry. Birdseye patented other machinery which cooled even more quickly. In 1927, he patented the multi-plate freezing machine which was used as the basis for freezing food for several decades (Richards, 2012). After much persuasion of E.F. Hutton by Merriweather, Birdseye agreed to sell his company and patents for $22 million (approximately $335 million in 2021 dollars) in 1929 to The Postum Company, which eventually created a line of frozen foods under The Birds Eye Company brand.

iii. Reinventionaries

Reinventionaries are firms that proactively recognize that their current lines of business might wane and profits erode due to the upcoming industrial revolution. This erosion of these firms existing business might be because the products and services they are offering are in decline, or are likely to be disrupted by advancements in technology, rendering their current product irrelevant and obsolete. These companies do certainly exercise a degree of risk in transforming themselves and self-disrupting their business models. Some firms make dramatic shifts in recognizing the impacts of changes ahead, others make subtle shifts in their models. For some, the change in their product offerings are a hard pivot, while some others make a more methodical, gradual, or precise shift, developing their new products while decreasing emphasis on their main product.

Depending on the industry, the market, and the urgency of the change, firms in this category generally seek to make business model changes – regardless of how dramatic these changes are – from a position of strength. In other words, the changes that these companies make are in response to recognizing external stimuli, opportunities or threats, and range the spectrum of risk taking. Some companies make major shifts, expanding into new industries, taking risks to expand into unfamiliar terrain to capitalize on opportunities and minimize risks in their current business models, others take measured and controlled risks. What is notable is that there is no guarantee that these firms will be successful amidst the change, but they recognize that inaction is not an option. As the adage made famous by Ret.

General Eric Shinseki that was mentioned earlier in this book goes, these firms operate by the tenet "If you don't like change, you'll like irrelevance even less."

There are several companies that reinvented themselves in response to global trends brought upon at the start of, or at the maturing of an industrial revolution. Notable amongst this list are two technology companies that we will explore further along in this book, Apple Computer and Netflix. Apple Computer reinvented themselves with Steve Jobs in his second stint at the helm of the organization that he founded. Facing stiff competition from Microsoft (and Bill Gates/Steve Ballmer) through the 1990s, where PCs running Microsoft Windows operating systems became ubiquitous in homes and offices, Steve Jobs made the decision to continue development of the Macintosh line of computers, building on its cult-like following, while pivoting Apple Computer into consumer electronics. Establishing a self-contained ecosystem with the iTunes Store and farther along, the App Store, Steve Jobs and Apple Computer reinvented themselves to emerge from flirting with obsolescence to dominate consumer electronics for years to come. The iPhone accelerated the adoption of the smartphone and gave birth to a whole new industry – the industry of apps.

Imagine our lives without smartphone apps today – whether you are an Apple aficionado and an iPhone/iOS user, or use an Android-based smartphone, with the same apps available on the Google Play store, we are fully reliant on shopping, transacting, and getting information and entertainment courtesy hundreds upon thousands of these apps. That has been a revolution onto itself, and this revolution commenced with Apple's reinventing themselves. The other example we will delve into will be that of Netflix. Netflix for the first several years of their existence used to be a DVD/Blu-Ray Disc subscription model. A subscriber would travel to Netflix' website, select their DVDs, and these would be mailed to the subscriber via the United States Postal Service. Their decision to "grandfather" and ultimately "sunset" their DVD business in favor of 100% online streaming was a total reinvention of a relatively young company. They caught lighting in a bottle, and Netflix today is one of the most successful businesses in the history of the planet.

iv. Responsive and Reactionaries

Responsive and Reactionary companies find themselves occupying the same grouping. This is primarily due to the fact that in both subgroupings, the underpinning is the same – these firms seek to make changes reactively - as a result of an external force/set of forces. Their actions are as a result of responding to external changes or reacting to them. WHEN these firms decide to actually go ahead and execute on the changes, that is, how swiftly these firms take action, is what determines if they are responsive companies or reactionary ones.

Responsive companies see opportunity arising from the change, or at the least recognize risks to their operating models and business strategies quite early on after impacts of these macro changes start to be experienced and felt. While they certainly cannot be considered to be visionaries or capitalizers (since those types of companies are proactive in advance of changes being experienced or felt), the "responsives" are early adopters of adaptation to changing circumstances around them. The core difference between responsive companies and reinventioners is the fact that responsive companies take action reactively, while reinventioners take action proactively. Even though their actions are reactive in nature, responsives take these actions early enough when feeling macro changes that they straddle the line between reinventioners and reactionaries. Their shifts in product strategies, business plans, or strategies are thought of being made from a position of strength despite that they might need to effect change under some level of duress. There is inherent risk involved for these firms to undertake making these business changes, but they recognize the rewards of making these business changes or realize the risks of not doing so. Responsive companies are thought of as flexible and adaptable, conforming and adapting to the needs of the marketplace and responding to their customers' expectations. These firms develop a corporate culture that is designed for uncertainty.

Domino's Pizza is a great example of an organization that found itself very quickly in deep trouble and consciously decided to totally overhaul their business. Domino's Pizza, despite having placed emphasis on maintaining a positive brand image, was struggling around the period of the Great Recession in 2008 as their stock price had hit an all-time low at $3 a share. Exacerbating their problems was that in 2009, two employees posted a video on YouTube that showed them tampering with food. Worst yet was the fact that customers by-and-large "thought the pizza was lousy" (Profit&, 2019). The way they went about it was astonishing for the approach they took. Acknowledging that their consumers had been abandoning them due to poor tasting products and less than stellar service, the company aggressively promoted their own recognition that they had lost touch with who they aspired to be as an organization. Their CEO at the time, Patrick Doyle, "appeared in ads with Domino's workers reading blunt reviews: 'worst pizza I've ever had,' 'sauce tastes like ketchup' and 'the crust tastes like cardboard.'" (Meyersohn, 2018). The candor and transparency the company projected during this transformational period helped them blunt a feeling of "imminent demise", buying them goodwill with consumers and investors as they sought to steady the ship.

Predicated on digitization and seeking to reestablish themselves as a technologically progressive and innovative company, Domino's implemented changes that impacted every single aspect of their value chain – including their core product, pizza. In addition to overhauling their 49-year old pizza recipe, "the brand implemented new technology to support the chance. A new custom delivery vehicle with a heating oven was introduced, dubbed the DXP, which acted as a form of advertisement despite only 150 being on the road at the time. The brand ramped up its digital efforts as well to meet consumer demand. Text messages, Alexa, Google Home, Twitter, Facebook, Smart TVs - they're all methods used by consumers to order a pizza. Domino's leveraged the wealth of consumer data through its custom operating system. This helped keep the transaction costs low and provided Domino's with insights about its customers. Then there's also the case of developing loyalty programs and introducing special offers to continue to drive up sales. Despite the successful change, it hasn't stopped there. The brand has also tested drone and robot delivery - even partnering with Ford on self-driving options." (Profit&, 2019). Domino's digital-first approach has earned them praise for being easy to use, customer-friendly, and garnered them the support of a younger customer base that is generationally skewed towards transacting 100% using their smartphones.

Reactionary companies are those that shift their business strategies and corporate plans as a result of feeling mild to moderate negative impacts from changes around them. In this scenario, these firms usually find themselves challenged by their existing products and services being challenged on account of being potentially rendered obsolete due to technological advancements, or an "upstart" competitor/a disruptor having successfully figured out how to capitalize on these technological advancements so much so that they threaten to upend the established corporate competitive landscape. Reactionary firms usually have to make business changes under duress and these business changes are typically associated with a potential significant amount of restructuring, job disruptions, and can seem quite chaotic depending on the urgency of the need to change.

The distinction between responsive companies and reactionary ones is how soon after macro changes start being experienced that they reactively enact business strategy changes. Nokia is an example of a reactionary organization that we will revisit later in this book. Nokia, a Finnish company, dominated the mobile phone market in the late 1990s through the mid-2000s. According to an article appearing in the New Yorker, "Historically, after all, Nokia had been a surprisingly adaptive company, moving in and out of many different businesses—paper, electricity, rubber galoshes. Recently, it successfully reinvented itself again. For years, the company had been a conglomerate, with a number of disparate businesses operating under the Nokia umbrella; in the early nineteen-nineties, anticipating the rise of cell phones, executives got rid of everything but the telecom business.

Even more strikingly, Nokia was hardly a technological laggard—on the contrary, it came up with its first smartphone back in 1996, and built a prototype of a touch-screen, Internet-enabled phone at the end of the nineties. It also spent enormous amounts of money on research and development. What it was unable to do, though, was translate all that R. & D. spending into products that people actually wanted to buy." (Surowiecki, 2013). A hardware organization at its core, Nokia failed to recognize the equitable importance of software, a concept that Apple and the iPhone revolutionized with the Apple App Store and iTunes Music.

Nokia leadership, fearing that they might alienate their current used base if they changed too much resisted leading a dramatic change in user experience. They also overestimated the strength of their brand and market share, and resting on past laurels, considered themselves to be impervious to advances by Apple and Google (with Android). By the time they decided to compete with Apple and Android, it was far too late. The aforementioned New Yorker article summarizes the fall of Nokia, and others like it, the best: "The high-tech era has taught people to expect constant innovation; when companies fall behind, consumers are quick to punish them. Late and inadequate: for Nokia, it was a deadly combination." (Surowiecki, 2013).

v. Laggards

Laggards are organizations that effect changes to their business plans and shift their strategies extremely late in response to macro changes. These organizations usually make changes under extreme duress, facing eroding profits and struggling to maintain relevance. Organizations that are less well-funded and do not have access to deep cash reserves typically risk going out of business as a direct result of how delayed their shifts in strategies occur. By this time, these organizations would have lost market share, face customer and employee attrition, and undergo significant turbulence. These periods of effecting reactive changes are often quite tumultuous for the organization, and consequently, its employees. Companies might face stiff competitive pressures as a result of better-funded competitors, or those firms that have implemented business changes much earlier and from a better position to be able to do.

These companies make changes from a position of relative weakness, and these changes are intended to stabilize the firm, put them on a stable footing, and simply survive. Growth, profits, and employee development usually take a backseat to staying afloat long enough to be acquired by another organization, or for their change in business strategies to start bearing fruit.

The Eastman Kodak Company has the unfortunate distinction of being recognized as a laggard. Kodak, founded in 1892, at its root, was considered to be an innovative company that leveraged technology heavily to utterly dominate the photographic film market for most of the 20th century. The fact that their tagline - the "Kodak moment" - made its way into common lexicon in order to describe any event that warranted being preserved for creating personal memories, is an indication of how dominantly the company was associated with all things photography and capturing and memorializing events big and small. However Kodak went from their commanding position in the industry to filing for Chapter 11 Bankruptcy in 2012. The company failed to recognize the rapid pace of digitization and the digital era that was upon them until it was far too late. Having lost significant market share to Fuji Film in the digital photography market, Kodak continued to face steep declines in their core product line – film-based photography – until smartphones and the ability for humans to take digital photos for their "Kodak moment" via their phones served as the metaphorical "final nail in the coffin".

The irony in this saga is that Kodak themselves had first invented the digital camera in 1975 thanks to Kodak engineer Steve Sasson. Since it was filmless photography and photography film formed the bulk of their core business, Kodak leadership told Mr. Sasson to not tell anyone about his invention. They failed to view digital photography as what would become the new normal. Considered to be disruptive technology at the time, despite inventing the world's first consumer digital camera, Kodak leadership did rejected a request to launch, market, or sell this product to protect eroding sales of traditional film. Where Kodak failed, Fuji Film did not. Fuji heavily invested in digital photography and overtake Kodak as the dominant player in the photography market. Having sold many of its patents to companies such as Apple, Google, Microsoft, Samsung, Adobe, and HTC, Kodak emerged from bankruptcy protection on September 3rd 2013. The company operates via five main business divisions today - Print Systems, Enterprise Inkjet Systems, Micro 3D Printing and Packaging, Software and Solutions, and Consumer and Film – at a mere fraction of the dominant market position it once occupied.

vi. Obsoletioners

And finally, we have the most unfortunate category of companies in this grouping list. These category is comprised of companies that went obsolete due to failing to recognize the opportunities and risks emerging from macro changes resulting from an industrial revolution. Not only do these companies fail to proactively recognize the telltale markers of an upcoming revolutionary change, they ignore the harbingers of these changes for far too long to even make reactive changes to their own business in order to survive, much less thrive amidst macro changes.

The corporate world is littered with companies that went obsolete due to their failure to respond to change. Sometimes, firms that become obsolete have little to no option to change. They simply specialize in products and services that no longer are relevant to the era, or are replaced by a much more desired alternative. We broached the topic of horse breeders and horse buggy manufacturers earlier on in this book. Even the most progressive, change-embracing horse buggy manufacturers could have done little to control their main product lines displacement due to automobiles. Had they even wanted to change their business plans or strategies, what could they have evolved their companies to do? It is for sure that some of them could have repurposed their assembly plants to help create automobile chassis, or other parts for automobiles, but imaginably any company with this foresight would be greatly limited in being able to envision and explore options. How about the horse breeders in this scenario? Bred as the primary mode of transportation for hundreds of years, horses were suddenly rendered obsolete as a mode of getting from one place to another. There could have possibly been very limited options when a horse breeder had to upend their own business model at the turn of the 20th century.

That said, it is also true that by-and-large, companies that go obsolete are ones that are beyond change averse, they are change resistant, and lack the deep financial resources to be able to weather the most disruptive of macro changes around them. They run out of cash and they run of time and through this entire period either do not make the right changes or changes with some substantive depth. Their demise is not inevitable and they have opportunities – at times multiple opportunities - to course correct. If you step back and look at this very list, companies that become obsolete trip over every single failsafe in this continuum. They fail at driving change as visionaries, they fail at being proactive and reinvent themselves, they fail at being reactive and responding in a reasonable period of time, and finally even if they are laggards, they lack the wherewithal to make the changes they needed to make quick enough and effectively enough.

I talk about of Blockbuster quite a bit in my speeches on digitization, and also referred to them as a case study (of what not to do) in a TED Talk I had the privilege of giving in September of 2019. I see Blockbuster as a classic example of a firm doubling down on their inability to change and their incapability of resting on their own laurels. One of the most recognizable stores at the corner of strip malls across the United States and internationally, Blockbuster was iconic in that it was also an integral part of most people's lives – on how they sought entertainment. It required some advance planning to be able to get to a store, understand what you wanted to watch, reconcile your preferences with others in your household (since most homes used to have one or two television sets/VCRs/DVD players in their homes), bring home physical media, reserving copies of movies that you wanted to watch but had been rented out, and then returning the media in time to avoid being charged outrageous late fees.

Blockbuster was a dominant force in the video rental market for decades, operating at its peak in 2004. They had successfully navigated a steady transition from VHS tapes to DVDs and eventually Blu-Ray Discs in addition to DVDs. Blockbuster failed on multiple levels, but their two primary failures were that they failed to adapt to a marketplace that delivered entertainment to customers rather than having customers come to them, and much like the example of Kodak failing to move away from physical photography film to digital photography, Blockbuster also failed in anticipating a shift from physical media to streaming entertainment. While the rise of Netflix is most frequently attributed as the cause for Blockbuster's demise, it is only part of the story. Blockbuster was much better funded and in a much more financially position to compete with

Netflix in the early days of the latter. When Netflix started shipping DVDs directly to their customer's homes as a subscription service, Blockbuster had an opportunity to partner with Netflix.

Given their strengths, it was foreseeable that over time, Blockbuster might have been able to potentially ultimately acquire Netflix. In 2000, Netflix founder Reed Hastings had proposed a partnership to Blockbuster. Hastings proposal would have had Blockbuster advertise Netflix in their stores, and reciprocally, Netflix would serve as Blockbuster's online arm. Deriding Netflix's business model as a "niche business", Blockbuster declined Netflix's proposal. When Netflix decided to reinvent themselves and shift from mailing physical media (DVDs/Blu-Ray Discs) to their customers to becoming a subscription streaming service, it presented yet another opportunity for Blockbuster to partner or pursue an acquisition of Netflix. Once again dismissing streaming as a niche industry and skeptical that Netflix's business model would work, Blockbuster passed on another opportunity to work with Netflix. Either opportunities would have saved the company from obsolescence. Blockbuster filed for Chapter 11 bankruptcy in 2010, steadily shutting down all their stores over the next few years, while Netflix rapidly became a global behemoth, valued at $28 billion dollar company as of 2021.

Those Who Don't Learn From History

Even though each industrial revolution is different; distinct in the level and type of disruption that it brings, the patterns of how industries and firms react in response to the avalanche of changes brought about by each industrial age follows familiar patterns. The variability in how each firm within an industry navigates through an industrial revolution determines winners and losers – in terms of corporations – for that specific era. However, workplaces are meaningless without workers. Workers - real people – with skills, talents, their own strengths and their own opportunities for growth. Workers can become the collateral damage in the wars of a company that does not exhibit the foresight to adopt and adapt to this change. Without understanding how your particular company might react, it is challenging – not impossible – to anticipate the trends within your industry holistically.

For instance, if you work for Acme Company and realize that the corporate culture at Acme Company is one where they are firmly entrenched in the "Laggards" category, you might want to explore new opportunities for employment (with the associated growth and development) in Company ABC within your industry because you see them as an adaptable company. You might realize that your career prospects with Company ABC, which might be in the "Responsive" or "Reinventionaries" categories, are much brighter. You might face career stagnation with your currently employer and your skills might atrophy while you Acme Company is stuck "on neutral" and fail to comprehend the nature and extent of changes that are all around them. Therefore, the journey to take control of your own destiny – to be able to truly "**thurvive**" in the "Fourth Industrial Revolution" – it is absolutely vital to understand what category your current firm is in. It is also vital to understand whether your current industry is dominated by companies in the "Laggard" or "Reactionaries" categories. If this is truly the case, you might be better off exploring career opportunities in an entirely new industry, and the sooner you make this switch, the better.

For example, think of whether you were gainfully employed as a salesperson selling horses for horse buggies at the turn of the 20th century. This is a profession that you have become an expert at because you've been doing it for twenty-five years. Regardless of if you worked for "Acme Company Horse Retailers" (no pun intended) or "Company ABC Horse Retailers" at the time, if you are personally able to envision the displacement of horse and buggies by automobiles, you can also be quite certain that your entire industry – not just your current employer – is unlikely to make shifts to be able to survive. In this scenario, switching companies within the same industry will not do. It will not yield you any benefit. This is when you will need to make the conscious (and possibly scary) decision to shift into an entirely new industry. Maybe you learn everything about these newfangled things called automobiles and become a salesperson for automobiles. You can segue your experience in the transportation industry as the basis for effecting this change.

Chapter Nineteen – Thurvival Technique 4: Develop Your Human Skills

It sounds like a paradox, but **to succeed digitally**, to "**thurvive**" amidst all this technological change, you have to **be more human**.

We've seen example after example of individuals who were able to learn skills and execute on what they had learned throughout this book. From the Polgar sisters and Josh Waitzkin to the Silicon Valley entrepreneurs who created everything from the smartphone to the internet seemingly out of thin air, these individuals were taught, or taught themselves, teachable and learnable skills. Anyone can learn anything. Some things might be more challenging to learn. Not all of us might have the same level of aptitude, inclinations, and predisposed proclivities to pursue a particular passion (say that rapidly five times). However, despite how you feel about a particular skill, be it playing an instrument if you're not musically inclined, or learning martial arts even if you're not particularly coordinated, the immutable fact remains that anyone – each and every one of us – can learn any skill. No skills are easy when you start.

Even an expert was once a beginner, and no expert was good when they started. As the saying goes, 'you don't have to be good to start, but you have to start to be good'. With time, tenacity, patience, practice, and persistence, even the most "unskilled" can become skilled. These are what I call the "hard" skills. Hard does NOT mean difficult. Hard means that there is definition to the skill, and structured training is available for anyone – regardless of skill level – to operate within, learn from, and get quite proficient at the skill. Technology – even if you an IT professional – is a hard skill. It needs to be taught or self-taught and has a defined structure to learn within. It is not particularly difficult, but can seem intimidating or challenging to those who aren't "technically inclined". Just as reading music notes would be intimidating to someone who isn't musically inclined, becoming learned in the field of technology – even at the most cursory levels – can come across as an uphill task. But I can assure you, technology – learning technology – is not difficult.

In an age dominated by rapid technological change, the value of soft skills has become more important than ever before. As our world becomes increasingly automated and digitized, the ability to connect, communicate, and empathize with others sets us apart as uniquely human. Soft skills, often referred to as "people skills" or "emotional intelligence," are interpersonal abilities that allow us to effectively navigate and succeed in our personal and professional lives. Soft skills encompass a wide range of personal attributes, including but not limited to, communication, emotional intelligence, adaptability, teamwork, leadership, and problem-solving. While hard skills, such as technical knowledge and expertise, have traditionally been the focus of attention, soft skills provide the foundation upon which successful human interaction and collaboration are built. As technology continues to automate routine tasks, it is the soft skills that will enable individuals to navigate complex challenges, build meaningful relationships, and thrive in the evolving workforce. In the digital landscape, where information cascades with unruly fervor, the ability to communicate effectively becomes an art form of paramount importance. Through eloquence, empathy, and active engagement, we can bridge the gap between humans and technology, ensuring that our ideas are understood, our intentions are conveyed, and our impact is amplified amidst the digital symphony.

The Softest Skills are the Hardest Skills.

While there are a wide range of programs that you can avail of to become more technically astute in order to "**thurvive**" in the Fourth Industrial Revolution, your **SOFT SKILLS** are going to be just as crucial for "**thurvival**" as your "hard" skills. Technology and AI can automate (and make exponentially more efficient) the most repetitive, mundane, and rote tasks. What technology and AI will find difficult to replicate are the quintessential, innate human skills. Leadership, ethics, morals, values, sound judgement, communication, collaboration, community-building, organizing, etc. are inherent human qualities that cannot easily be automated. At its most basic, soft skills in the 21st century are about retaining and nurturing the human touch. As technology becomes increasingly sophisticated, certain human qualities cannot be replicated by machines, and soft skills, such as empathy, emotional intelligence, and effective communication, allow individuals to connect with others on a deeper level. In a world where technology is poised to replace most routine tasks, the ability to forge meaningful relationships and understand others' perspectives becomes a differentiating factor. The human touch has an innate value that can never be fully replaced by algorithms or automation. These fundamental soft skills however are often not prioritized as much as learning hard skills are, and are often downplayed in favor of hard skills development.

This trend is even more pronounced in leaders within skilled trades such as information technology, actuarial sciences, as well as specialized fields like sales. Everyone can be a salesperson, but not everyone can sell. IT and actuarial type vocations are heavily centered on the technical aspects of the craft, and do not historically focus on soft skills and soft skills development. The stereotypes are not too far from the truth. Leadership in these fields requires a bedrock of knowledge and a breadth of expertise for certain, but success in these fields, in a leadership role, demands that the leader thinks of their specialized vocation second, and their ability to translate and be a bridge between their people, their skilled technical craft, and the outside world first. Leading these organizations, or leading organizations that comprise the skilled trades, and the "type A" personalities you may stereotypically expect in a sales team, requires a very different type of a multifaceted leadership approach. Technical trades are notorious for taking the most skilled, our highest performing technical professionals, an individual contributor, and promoting them into a leadership role by making them a "manager". With limited experience in honing their soft skills, and significant prior experience succeeding in individual skilled contributions, these newly minted leaders can struggle. They will undoubtedly develop their management styles and build their experiences over time, however a good number either pine for, or act upon, returning to their previous roles as individual contributors / senior technical professionals.

Soft skills development is further challenged by several factors that are unique to the third decade of the 21st century. In contrast to just a generation or two before them, Generation Z and the younger age spectrum of Millennials prefer to communicate via their smartphones versus engaging in direct dialog with one another. Texting, and most recently, communicating via apps such as Snapchat, have taken precedence. Generation X and the older age band of the Millennials often wax philosophical about an era where all they were told by their parents was to "go out and come home before it gets dark". Kids of that generation played and interacted with each other – in person – for hours, developing their social skills in the process. Perhaps the concerns of how the latest generations are interacting in society are overblown. We quite simply don't know how this will turn out. We do not have any data, nor is there a precedence for this way of human interaction. Another significant event that exerted pressure on soft skills development has been the COVID-19 pandemic.

Workers entering the workforce today, those that belong to Generation Z, prefer to work remote, or in a hybrid environment, where they are in their home offices more than in their office buildings. While this has been a boon for striking an effective work-life balance and productivity, it can have the effect of stunting soft skills. It becomes more difficult to lead individuals in such environments, and challenging to inspire your corporate culture and values, when the engagement between workers can be relegated to a video conference. Unless companies are intentional and focused on fostering their corporate cultures and bringing their employees together on a periodic manner, this can be adverse to soft skills development – especially for those new workers who have never worked before, and potentially have never set foot in an office campus. An additional concern is that as the Boomer generation approaches retirement age, the institutional knowledge and corporate wisdom that used to be passed along from generation to generation in a mentor and mentee relationship – a relationship that can ideally be fostered when there is in-person engagement – will be greatly diluted. This also results in new workers not being able to develop and nurture their soft skills.

While technical skills and knowledge remain valuable, they are no longer sufficient in isolation. As automation and artificial intelligence continue to automate routine tasks, the qualities that make us uniquely human are becoming increasingly prized. Soft skills such as communication, empathy, adaptability, critical thinking, creativity, and collaboration are now more crucial than ever before. These skills enable us to build meaningful relationships, solve complex problems, embrace change, and adapt to new environments.

Developing Your Soft Skills

A section on developing soft skills is where most individuals employed in skilled trades that rely more on NOT having to interact with others usually stop paying attention. However, if you are currently in a skilled trade such as IT, or planning to be, I implore you to not ignore investing in cultivating your soft skills. Your code can be authored or quality checked by an AI, but the good judgement of verifying that the AI is unbiased and explainable belongs to you. Communicating with others across your organization is a fundamental expectation of working as part of a team.

Leading others might not be your cup of tea, but leadership does not have to involve administratively managing them – even technical leadership IS leadership. You NEED soft skills.

Developing soft skills requires conscious effort and continuous practice. Here are a dozen strategies we can employ to nurture our soft skills in the midst of technological change:

1. Start with a Self-Assessment for Self-Awareness:

Start by understanding your strengths, weaknesses, values, and emotions. A personal SWOT (Strengths, Weaknesses, Opportunities, and Threats) matrix is a wonderful tool for you to plot out where your opportunities for improvement reside. These opportunities could be in intra-team communication, style of interactions, public speaking, presenting skills, etc. Once you have a baseline honest self-assessment of your opportunities, reflect on your interactions with others and seek feedback to gain insights into your communication style, emotional intelligence, and areas for improvement. This will allow you to build a personalized roadmap of sorts on how to get from point A to point B. It is absolutely ok to seek help – from a mentor, from your peers, from your superiors in your organization – or even formal training, or group training (such as Toastmasters for bettering your public speaking skills). This introspection is the foundation for personal growth and development.

2. Practice Active Listening:

Our world today is filled with distractions and information overload. The concept of active listening has consequently become a rare skill. It is expected that the information overload is only going to exponentially increase in the next several decades. This is where you can develop your soft skills and practice listening attentively, seeking to understand others' perspectives, and valuing their contributions. By doing so, you enhance your ability to collaborate, build rapport, and establish meaningful connections. People will appreciate having an engaged and captive audience, and allow you to strike an emotional chord with your coworkers. Relationships – at work or otherwise – are predicated on this simple, yet often esoteric concept. Empathy is a foundational soft skill in the digital age. Cultivating empathy involves actively putting yourself in someone else's shoes, understanding their emotions, and demonstrating compassion. Practice empathetic active listening by genuinely focusing on the speaker, seeking to understand their perspective, and responding with empathy. Engage in conversations with colleagues, friends, and family members, practicing empathetic communication to build stronger connections and foster understanding.

3. Practice Effective Written Communication:

Clear and concise communication is vital for conveying ideas, building relationships, and resolving conflicts. Invest in developing your verbal and written communication skills. There are numerous ways to go about achieving this – from online courses to LinkedIn Learning – there is no shortage of courses that focus on teaching improved communication skills. Written communication, especially email, can portend tonality where none is implied. In other words, emails can come across as open to interpretation, and this can often be the root cause of misunderstandings (and miscommunication). Effective communication requires that we pay special attention to tone, seek to convey a level of empathy, and develop the ability to tailor our messages to different audiences. Finally, written email communications are more formal means of information exchange than SMS text messaging. Avoid using shorthand or making an email stylistically appear as if it were a text message.

4. Develop Effective Verbal Communication:

Most people are as averse to the concepts of death, divorce, and moving homes as they are to the concept of public speaking. Public speaking does not have to be from a big stage in front of a large crowd of coworkers. I can also feel a third of readers tense up as soon as the words "public speaking" were featured as part of effective verbal communication. Speaking in a small team – a group of coworkers or superiors, a Scrum Team, Agile Scrum ceremonies, etc. – are also forms of public speaking.

Most people fear public speaking because they're afraid of how people will perceive them – and that they're going to be judged by their coworkers. It is absolutely true that when people are listening to you speak, they are rooting for you to succeed. Unless you are engaged in a naturally hostile dialog, which requires a very different set of communication skills to deescalate and diffuse situations so that constructive communication can happen, most of your listeners want you to successfully relay your points. Public speaking is also akin to building muscle by exercising. The more you do it, the more confident you will become, and the more confident you will become, the better at it you will get. Most people shy away from taking the first step, or retreat if one of two presentations don't go well for them. But this is where patience and persistence are key. You have to be tenacious at getting better at public speaking. The best way to do this is to embrace opportunities to improve your public speaking, storytelling, and negotiation abilities.

For the next generation of leaders, the ability to synthesize a vast amount of data and tell a story – either to your teams, your customers, or your superiors – is going to be an invaluable skill. Soft skills endow us with the ability to distill complex concepts into digestible narratives, to craft compelling messages that resonate, and to listen with unwavering attention. People don't remember facts and figures by themselves, but they certainly will recollect facts and figures if they're enrobed in a compelling story. Soft skills development for effective verbal communication will require us to enhance our ability to convey ideas, actively listen, and engage in constructive dialogue. There is no magic potion for becoming an effective verbal communicator. You have to put in the work by practicing clarity, conciseness, and empathy in your interactions, both in-person and through digital platforms.

5. Nurture Emotional Intelligence:

Emotional intelligence requires us to cultivate the innate human traits of empathy, understanding, self-control, and self-regulation. Emotional intelligence encompasses our ability to understand, manage, and express emotions, as well as to perceive and empathize with the emotions of others. Empathy, self-awareness, and the ability to regulate one's emotions are at the core of emotional intelligence. We can develop our empathetic skills by actively seeking to understand others' emotions and experiences.

While technology can analyze data and provide logical solutions, it lacks the inherent emotional understanding and empathy that humans possess. The ability to connect on an emotional level, provide support, and navigate complex interpersonal dynamics is a uniquely human trait that cannot be easily replicated by any technology, including AI. The tapestry of emotional intelligence enables effective collaboration, leadership, and conflict resolution. Emotional intelligence is especially crucial for effective communication and collaboration under times of dispute and duress. Emotional intelligence can help you diffuse and deescalate conflicts and find a path forward in the midst of emotionally heightened situations. As a learned skill, we should practice emotional regulation in order to navigate conflicts constructively and handle stress effectively. Emotional intelligence allows us to embrace diversity and cultural sensitivity, allowing for more inclusive and collaborative environments.

6. Develop Adaptability and Resilience:

We have discussed adaptability and resilience as foundational elements for us to successfully navigate the Fourth Industrial Revolution and set ourselves up for **thurvival**. Adaptability and resilience, although they are innate attributes and personality traits, can also be key soft skills that anyone can build and develop. Adaptability and resilience requires each of us to embrace change and learn to thrive in dynamic environments. We have to invest in building our capacity to adjust to new technologies, processes, and roles, seek opportunities to step out of our comfort zones, take on new challenges, and view setbacks as learning experiences. In a rapidly changing world, adaptability and flexibility are invaluable skills. While technology can be reprogrammed to perform new tasks, it lacks the innate ability to quickly adapt to new circumstances, learn from experience, and navigate unforeseen challenges. We possess the cognitive flexibility and resilience to embrace change, acquire new skills, and adjust our approach in dynamic environments, making us invaluable in situations that require creativity, problem-solving, and adaptability.

Through the Fourth Industrial Revolution, it is only by cultivating adaptability and resilience that we can remain agile in the face of uncertainty and thrive in evolving environments. In an era characterized by rapid innovation and perpetual change, the ability to adapt is a prized virtue. Soft skills imbue us with the nimbleness of thought, the agility of action, and the capacity to navigate the uncharted waters of digital disruption. Through adaptability, we can embrace new technologies, pivot seamlessly amidst evolving landscapes, and harness the winds of change to fuel our growth. Soft skills such as flexibility, creativity, and problem-solving enable us to navigate uncertain and ever-changing circumstances. While technical skills may become obsolete over time, soft skills provide the foundation for continuous learning, growth, and professional evolution.

7. Develop your Critical Thinking, Judgement, and Complex Problem-Solving skills:

Critical thinking involves the ability to analyze, evaluate, and interpret information objectively. It encompasses logical reasoning, problem-solving, and making sound judgments based on incomplete or ambiguous data. While technology can process vast amounts of data and provide insights, it lacks the contextual understanding and subjective judgment that humans bring to the table. Technology excels at making decisions based on predefined rules and algorithms. However, when faced with unstructured or unfamiliar situations, humans possess the capacity to consider multiple factors, weigh risks and benefits, and make complex decisions that balance various conflicting interests. Humans can integrate diverse information sources, evaluate long-term consequences, and incorporate ethical considerations, enabling them to make nuanced judgments that consider the broader impact on individuals and society.

Human critical thinking involves intuition, ethics, and the ability to navigate complex moral dilemmas, making it an essential skill that technology – including AI - struggles to replicate. In the face of intricate, ill-defined problems, humans excel at employing critical thinking and problem-solving skills. Humans possess the ability to analyze complex situations, gather information from diverse sources, and make decisions based on intuition and judgment. Technology, such as AI algorithms are limited by their programmed rules and lack the flexibility and contextual understanding necessary to tackle multifaceted problems that require creativity and adaptability.

Soft skills such as critical thinking and problem solving empower us to think, well, critically, to explore uncharted avenues, and to transcend the confines of linear thought. Technology thinks in a linear fashion, humans do not. It is the interplay of technical acumen and the finesse of soft skills that unlocks the tapestry of creative problem-solving, propelling us towards pioneering solutions. Develop your ability to analyze complex situations, think critically, and make informed decisions. Practice creativity and innovation, seeking alternative perspectives and solutions to overcome challenges. Soft skills, like any other skill, require consistent practice and refinement. Engage in lifelong learning by attending workshops, reading books, taking courses, and seeking feedback from mentors or peers. Embrace new experiences and challenges that stretch your abilities and provide opportunities for growth. The march of technological progress often presents us with intricate problems that defy conventional solutions.

It is through the prism of soft skills that we can unleash our innate creative potential, fostering innovation and envisioning novel paths forward.

8. Actively Practice Collaboration and Teamwork:

"No person is an island" as they saying goes. "If you want to go fast, go alone, but if you want to go far, go together" as an African proverb says. Just as exists the modern challenge of communications and corporate culture retention in a hybrid world, teamwork and collaboration – aspects built on effective communication – might also be challenged devoid of intentional focus on nurturing and fostering the same. **Thurvival** in the 21st century will require each of us to rely on our teams more than ever before. Remember that every person is on the same journey as you are, navigating through unchartered waters, and are wrestling with the same challenges, and are in pursuit of the same opportunities. The era will create plenty of opportunities for each of us, and of all the things you need to focus on, a sense of "winner takes all" competitive spirit, should not be one of them. It is within this environment that it is much more important to come together – and stay together – as a team, learning from one another, lifting each other help, and aiding each other in the quest for thurvival. A great team player is invaluable at any time, but a great team player is indispensable during this particular time in our history.

Foster your capacity to work effectively in diverse, potentially geographically distributed teams. We should embrace cooperation, active participation, and constructive conflict resolution. To develop this soft skill, we have to recognize and value the strengths of others, leveraging collective intelligence to achieve shared goals. Collaboration is crucial in today's interconnected world. We should foster teamwork by appreciating diverse perspectives, valuing teamwork, actively contributing to group efforts and developing skills in negotiation and compromise to create harmonious working relationships.

9. Develop Creativity and Innovation Skills:

Creativity is at the core of human ingenuity. Creativity, an intrinsically human trait, involves the generation of novel ideas, the ability to think outside the box, and the capacity to innovate in various domains. While machines excel at repetitive and rule-based tasks, they often struggle with creativity and innovation. While technology can help in the analysis of existing data and patterns, it often struggles with *true* creativity. Human creativity stems from our unique experiences, emotions, and imagination, enabling us to come up with groundbreaking ideas, solve complex problems, and drive innovation in ways that technology simply cannot match. Soft skills such as critical thinking, problem-solving, and lateral thinking are essential for generating fresh ideas, identifying opportunities, and adapting to new situations. The ability to think outside the box and approach challenges from different perspectives enables individuals to contribute unique insights and create innovative solutions that drive progress in the digital landscape.

10. Be a Passionate Advocate for Ethics and Morality:

Ethics and morality guide our actions, decisions, and interactions with others. While technology can be programmed with ethical guidelines, it lacks the capacity for moral reasoning and the ability to understand the nuances of ethical dilemmas, making it challenging to navigate complex ethical decisions without human input. Technology can adhere to programmed rules and principles, it lacks the intrinsic moral compass and ethical judgment that humans possess. We can navigate ethical dilemmas, understand the nuances of ethical reasoning, and consider the broader societal implications of our choices. The ability to balance conflicting values, make ethical judgments, and act with empathy and compassion makes human ethical decision-making a crucial skill in domains such as healthcare, law, and governance. Ethics and moral reasoning are integral to human decision making. We possess a sense of ethics that enables them to weigh the consequences of our actions, consider the impact on others, and make morally sound judgments. Being a passionate champion for ethics and morality within your organization will ensure that you continue to lean into what makes you intrinsically human, a soft skill that technology will unlikely be able to replicate.

11. Exercise your Leadership and Lead by Influence:

A leader is responsible for guiding and mentoring other people. When we think of leadership we automatically gravitate to leadership within a professional setting and establish a word association to business, to industry, or to the corporate world in general. Most of the time we conflate the terms "leader" and manager. Although often used synonymously, I fundamentally believe that there is a distinct difference between managers and leaders. It is my belief that the verbiage distinction is an important one to make. It is my belief that you manage things but you lead people. A leader is not necessarily just the person at your workplace, the one who provides you coaching and feedback, and perhaps does your annual performance review. Leadership is at once an incredible honor and an immense responsibility.

Each one of us at some time, in some way, has demonstrated our leadership abilities and is called upon to lead as situations and circumstances warrant. There are those who pursue leadership roles within their organizations, organically find themselves in such roles, or are recognized for being leaders or pioneers within their fields. At every point in our careers, each and every one of us, even those who do not serve in "management" roles, will serve as leaders to someone else. Whether this is in a formal setting, such as within the auspice of the corporate world, or in an informal or organic setting, such as within the context of our social or interpersonal relationships, each of us will, at some point, on someday, and in some way, lead others.

A leader has the ability to influence someone's life, to make an immense impact. As a leader, you might not realize, fully grasp the magnitude of impact you are having on your talent, both good and bad. This impact can be for a moment in time or trigger a chain of events that make a difference over someone's lifetime, professionally and personally. It is not something to be taken lightly, especially when this responsibility of leadership has been bestowed upon you in a corporate setting. Even if you might not be a leader in terms of managerial responsibilities within your firm, you might, at some point, serve as a technical subject matter expert to others, guiding them – and effectively leading them, as well as their growth and development.

Leadership is a vital soft skill in the age of digital transformation. We should actively seek opportunities to take on leadership roles, even in non-traditional settings. This could involve leading a project, mentoring colleagues, or spearheading initiatives within your organization or community. Leadership experiences help develop skills such as effective decision-making, influencing others, and fostering a positive work culture, all of which are valuable in the digital landscape. The Fourth Industrial Revolution is going to be in dire need of strategic, stoic leaders – empathetic coaches whose only priority is the nurturing and development of their people. Effective leadership requires a combination of interpersonal skills, strategic thinking, and the ability to inspire and motivate others. Leadership is a distinctly human attribute that encompasses vision, empathy, and the capacity to understand and align diverse perspectives.

Technology may assist in data-driven decision-making, but it cannot replicate the art of leadership and the ability to inspire and guide individuals and teams toward a shared purpose. Whether you serve as a "manager" or a technical leader, soft skills development in the aspect of leadership will equip us with the qualities that elevate us to the mantle of leadership – an inspiring vision, effective communication, empathetic guidance, and the ability to foster collaboration across diverse teams. By cultivating our leadership soft skills, we can empower ourselves to lead with integrity, influence with compassion, and navigate the intricate labyrinth of the digital landscape, guiding others towards success and growth.

12. Start Now, Start Today:

Amidst the vast, unchartered ocean of new technology, waves of digital change, and cresting AI advancements, only the essence of human connection that remains unparalleled. Soft skills empower us to forge authentic bonds, to communicate with finesse, and to understand the unspoken cadence of emotions. It is through effective collaboration, empathy, and interpersonal dexterity that we can transcend the barriers of technology and create meaningful connections that reverberate with resonance and depth. Soft skills – like the "hard" skills – are learned, cultivated, nurtured, and developed. In order to prepare for the tomorrow, we must start now and start today. There are a multitude of ways in which each of us can pursue growth and development in building our soft skills in order to "**thurvive**":

a. Training Programs and Workshops: Participate in workshops, seminars, and training programs that focus on enhancing specific soft skills that focus on interpersonal communication, emotional intelligence, leadership, and other relevant areas. Seek out opportunities to practice and receive feedback in a supportive environment. Soft skills, much like hard skills, can be developed and refined over time. Receiving constructive feedback is essential for growth and development of soft skills. We should proactively seek out feedback and be open to self-improvement to continually enhance our soft skills. By actively seeking feedback from colleagues, mentors, or supervisors, we can gain insights into our strengths and areas for improvement. Reflect on your interactions, identifying patterns and opportunities to refine your soft skills further. Incorporate the feedback received into your daily practice, adapting your communication style and approaches accordingly. Leverage the vast array of online courses, tutorials, and resources available to develop soft skills. Platforms such as MOOCs (Massive Open

Online Courses) and educational websites offer a wealth of materials on communication, emotional intelligence, leadership, etc.

b. Mentoring and Coaching: Engage with mentors or coaches who can provide guidance, challenge your assumptions, and help you develop your soft skills. Learn from their experiences and seek their insights on effective strategies for personal growth.

c. Putting into Practice: We should put our soft skills into practice in real-life situations. Seek out opportunities to collaborate, lead projects, and engage in activities that require effective communication, problem-solving, and teamwork.

d. Seek out Diverse Experiences: To develop soft skills, it is vital to expose ourselves to diverse experiences that challenge our existing perspectives and broaden our understanding of the world. We should involve ourselves in activities that require collaboration, such as group projects, volunteering, or joining professional organizations, and seek opportunities to work with people from different backgrounds, cultures, and disciplines. Such experiences will help to foster empathy, adaptability, and effective communication skills, enhancing our soft skills repertoire.

e. Embrace Cross-Disciplinary Learning: The digital age demands interdisciplinary collaboration and problem-solving. We should embrace cross-disciplinary learning by exploring fields outside our expertise. This could involve reading books, attending lectures, or taking courses on subjects such as psychology, human-centered design thinking, or business management. By understanding diverse disciplines, we can bridge the gap between technical knowledge and soft skills, enhancing our ability to collaborate and innovate in a digital world.

The Harmonious Dance of Humans and Technology

As technology continues to advance, the need for soft skills will only intensify. While automation may replace certain job functions, the uniquely human qualities of empathy, creativity, and complex problem-solving will remain irreplaceable. The ability to connect with others on a deep level, foster innovation, and adapt to evolving circumstances will be crucial in a rapidly changing world. Technological advancement has undeniably transformed the way we live and work. Automation and artificial intelligence have taken over repetitive tasks, making it imperative for humans to develop skills that are uniquely human. While technology can provide efficiency and speed, it cannot replicate the depth of emotional connection, creativity, and critical thinking that soft skills bring to the table. Moreover, the increasing reliance on technology has raised concerns about social isolation, decreased empathy, and diminished interpersonal communication. Therefore, nurturing soft skills is not only essential for professional success but also for maintaining our humanity and fostering social cohesion.

Developing soft skills is a lifelong journey that requires dedication, self-reflection, and practice. We need to embrace the challenges and opportunities that technological change presents and recognize that as humans, we possess the innate capacity to learn, grow, and adapt. By nurturing our soft skills, we can navigate the ever-evolving landscape with confidence, enrich our personal relationships, and shape a more harmonious and resilient future. As technology reshapes our world, soft skills are the essential differentiators that allow us to "**thurvive**" amidst rapid change. By consciously developing and nurturing these skills, we can not only enhance our professional prospects but also foster meaningful connections, promote social cohesion, and preserve our humanity in an ever-evolving technological landscape.

By cultivating effective communication, emotional intelligence, adaptability, critical thinking, creativity, and collaboration, we can harness our unique human abilities to navigate the complexities of the modern world. Through lifelong learning and a commitment to personal growth, we can embrace the opportunities presented by technological advancements and create a future where humans and machines coexist synergistically. This journey toward nurturing soft skills is both transformative and empowering, enabling us to thrive amidst the ever-changing landscape of the 21st century. As we embrace the coexistence of humans and technology in the workplace, it is these distinctly human skills that will continue to define our value, contribute to our collective progress, and shape a future where human workers and intelligent machines collaborate harmoniously for the betterment of society.

Chapter Twenty - Thurvive

The age of rapid technological change and digitization stands as a pivotal moment in human history. We are living through some truly remarkable times - a seminal moment in our history that brings with it a paradigm shift that demands our unwavering attention and proactive response. In the preceding chapters of this book, we embarked on a journey exploring the profound impact of the digital age on our lives and, more specifically, our professional endeavors.

We have witnessed how technology has permeated every facet of our existence, continues to reshape industries, is redefining jobs, and is challenging traditional notions of work. As we conclude our journey, delving into the realms of technological advancements, and the transformation of work, one truth resonates with profound clarity - embracing technology is not merely a choice, but an essential endeavor if we are to secure and future-proof our careers in this era of exponential change; a quintessential requirement if we are to "**thurvive**".

This Journey Ends, The Voyage Begins

As we reach the culmination of this book, we can look back on our journey of four steps in order to build "thurvival" skills for the Fourth Industrial Revolution. First, we sought to first "**Understand the Change**". This premier step in our journey was to recognize that technology-driven changes are an inevitability. We took a step into the future in this first step and took a peek at the most conservative likely look into our future. There will more than likely be technology introduced over the next few decades that is too fantastical to envision today. Next, we sought to "**Understand the Need to Have to Change**" - to understand that we do not have a choice to slow, stymie, or stop these technology-driven changes. We explored the past and sought to ground ourselves on the fact that while the past offers some prologue to the future, the increasing and accelerating pace of change over the past century is a harbinger on shape of things to come. Our third step was to "**Understand the Need to Want to Change**", a step intended to ensure that we are proactively addressing change readiness. In that section, we came back to the present after having explored the future and peeked at the past. We examined how the recent past is influencing our world today and what that might portend for our occupations over the next few decades. Our fourth and final step was to "**Understand How to be Ready for the Change**". This step urged us to adopt the outlined recommendations, building your own personal plan on the foundation of a continual learning mindset and keep exercising that continual learning mindset.

We learned the four techniques for "**thurvival**" in the Fourth Industrial Revolution. **Thurvival Technique 1. Rewire your thinking – Learn, Unlearn, Relearn** underscored why a pure continual learning mindset is inadequate for the era we are entering. We learned that the "**learn, unlearn, relearn**" mindset, predicated on the continual learning mindset is essential for us heading into the third decade of the 21st century. Technological progress is reshaping every aspect of our lives. From AI and automation to robotics, IoT, and virtual reality, the impact of these innovations is undeniable. The pace of change is accelerating, and embracing technology is no longer a choice - it is a necessity. It is essential for us to survive our own paradigms and equip ourselves to thrive in this rapidly evolving landscape. Lifelong learning has always been important, but it has now become an imperative. The knowledge and skills we acquire today may become obsolete tomorrow. To remain relevant and adaptable, we must cultivate a "**learn-relearn-unlearn**" mindset, one that unlocks our ability to navigate the uncertainties of technological change, seize new opportunities, and cultivate resilience in the face of disruption. The acceptance of technology and change can be daunting, and as it has for generations before, fear and resistance hold us back from fully embracing the possibilities that lie before us. It is crucial to confront these fears head-on, acknowledging them as natural responses to the unknown. By reframing our mindset, focusing on the potential benefits and opportunities, and seeking support and guidance, we can overcome these barriers and move forward with confidence.

With Thurvival Technique 2. Harness Technology, we learned about the need to be digitally literate. In an increasingly digital world, developing digital literacy – learning technology, how to apply it to our jobs, and how it will impact our careers and our industry - is essential. Digital literacy goes beyond basic technical skills; it encompasses the ability to critically evaluate and navigate the vast sea of information, and harness the power of digital tools for communication, collaboration, and problem-solving. By becoming digitally literate, we empower ourselves to engage with technology in a meaningful and responsible manner. We stand at the precipice of a future shaped by the intertwined forces of human potential and technological innovation. Leveraging technology is not a passive act, but a deliberate choice—a commitment to continuous learning, adaptation, and harnessing the infinite possibilities that lie within our grasp.

With Thurvival Technique 3. Understand your Industry and Company Trends, we learned that it is our responsibility to take full charge of our careers and understand where our industry broadly and our company specifically is headed in the age of digitization. Being digitally literate and technologically savvy is the base, but we need to understand the potential trajectory of our careers within the industry that we serve. In the nexus of human potential and technological innovation lies an incredible opportunity for synergy - an intricate dance where our inherent capabilities fuse harmoniously with the power of technology. By embracing technological tools, we amplify our creative capacities, supercharge our productivity, and unlock novel avenues for problem-solving and innovation. It is through this dynamic collaboration that we transcend the limitations of conventional roles and become architects of our own destiny, ready to navigate the uncharted terrains of the digital frontier. Harnessing technology as our ally, we can unlock new avenues for collaboration, allowing us to focus on higher-order thinking, complex problem-solving, and cultivating the uniquely human qualities that set us apart.

The digital age has reshaped the very foundations of work, rendering some traditional roles obsolete while birthing new opportunities. In this dynamic landscape, adaptability and a willingness to embrace technology are the compasses that guide us towards sustainable careers. By developing a growth mindset and actively seeking to upskill ourselves, we can remain agile and responsive to the changing demands of the market. Whether it is acquiring digital literacy, mastering data analysis, or understanding the intricacies of emerging technologies, our commitment to continuous learning ensures that we remain relevant and sought after in an era of rapid technological advancement.

Within the realm of technological progress lies a fertile ground for innovation and entrepreneurship. The democratization of knowledge and access to tools and resources have lowered barriers to entry, allowing individuals to turn their ideas into reality. Embracing technology empowers us to be creators, disruptors, and architects of change. By leveraging digital platforms, harnessing data insights, and leveraging emerging technologies, we can seize opportunities, build scalable businesses, and make a meaningful impact on the world.

With Thurvival Technique 4. Develop your Human Skills, we learned about the criticality of soft skills to complement the hard (technical) skills that we will need to do the jobs of today and tomorrow. The ability to adapt and be resilient in the face of change is a hallmark of success in the age of technology. Cultivating adaptability requires embracing a growth mindset, being open to new ideas, and continuously seeking out opportunities for growth and learning. Resilience allows us to bounce back from setbacks, learn from failures, and remain steadfast in the pursuit of our goals. Together, adaptability and resilience form the bedrock upon which we can navigate the turbulent waters of technological change. Individuals do not exist in isolation; we are part of broader social and organizational contexts. Fostering a culture of lifelong learning is crucial at all levels, from families and communities to educational institutions and workplaces.

While technology unlocks a world of possibilities, we must not lose sight of the importance of human connection. As we navigate the digital age, let us remember that behind the screens and algorithms lie real human beings with desires, emotions, and aspirations. Embracing technology should not come at the cost of neglecting our innate need for empathy, compassion, and genuine human interaction. By cultivating emotional intelligence, honing our communication skills, and fostering authentic connections, we infuse the digital landscape with the warmth and depth of our humanity.

Organizations must invest in learning and development programs, provide opportunities for skill-building and upskilling, and create environments that promote curiosity, collaboration, and innovation. By working collectively, we can create a society that embraces technology and lifelong learning, nurturing a brighter future for all. In the face of rapid technological change, it is vital to anchor ourselves in a purpose-driven mindset. Identifying our values, passions, and aspirations allows us to align our learning and technological adoption with a greater sense of purpose. By staying true to our core principles and goals, we can navigate the complexities of technological change with clarity and focus, contributing to a future that reflects our deepest values. We envision a future characterized by collaboration and innovation. Humans and technology working together, leveraging the unique strengths of each, to address the pressing challenges of our time. It is through this collaboration that we can create sustainable solutions, drive economic growth, and foster a society that values empathy, creativity, and lifelong learning.

Thurvive

The Fourth Industrial Revolution has brought about remarkable opportunities, enabling us to connect, collaborate, and create on a scale never seen before. From AI and automation to big data and the Internet of Things, these technological advancements have the potential to amplify our capabilities, augment our intelligence, and unlock unprecedented levels of productivity. However, with every transformative force comes a measure of uncertainty and apprehension. As humans, we possess an innate fear of the unknown, a reluctance to relinquish our familiar ways and venture into uncharted territories. But if history has taught us anything, it is that progress waits for no one. The digital age is upon us, and it is reshaping our professional landscape with astonishing speed. To flourish in this new era, we must overcome our apprehensions and proactively embrace technology as an ally rather than a foe. To thrive in this transformative era, we must proactively embrace technology, adopting a mindset of curiosity, adaptability, and continuous learning. By leveraging technology as a tool, we can transcend the limitations of the past, unleash our creative potential, and shape a future that amplifies our humanity.

Harnessing technology does not mean surrendering our humanity; rather, it presents an opportunity to elevate our humanity to new heights. By embracing technology, we can free ourselves from mundane and repetitive tasks, allowing our minds to focus on higher-order thinking, creativity, and innovation. Technology becomes a catalyst for our growth and development, enabling us to reach our fullest potential as professionals. This post-pandemic digital age has blurred the boundaries of traditional employment structures, ushering in the era of remote work, gig economies, and freelancing. While this shift may be disorienting, it offers unparalleled flexibility, empowering us to craft our own professional narratives and shape our work-life balance. The ability to work from anywhere, collaborate across borders, and tap into global talent pools opens up a world of possibilities for those willing to embrace the digital paradigm.

In this transformed professional landscape, continuous learning and adaptability become essential. Cultivating a "**learn-unlearn-relearn**" mindset is vital. Technology evolves at breakneck speed, rendering certain skills obsolete while creating new demands. Becoming lifelong learners, willing to acquire new knowledge, upskill, and reskill ourselves to remain relevant in an ever-changing world at a rapid pace has become especially vital. Technology has fostered connectivity and collaboration on a global scale. The digital age has dismantled geographical barriers, enabling professionals to forge partnerships, share ideas, and engage in cross-cultural exchanges effortlessly. This interconnectedness expands our horizons, introduces diverse perspectives, and nurtures innovation. To thrive in this networked world, we must actively cultivate our digital literacy, become adept at leveraging technology to connect, communicate, and create value. Soft skills, traditionally an area that has not received significant attention, will be a crucial determining factor for success in the Fourth Industrial Revolution. As we navigate these unchartered waters of the digital era, we must remain vigilant and mindful of its potential pitfalls. Issues such as privacy, data security, and ethical implications of emerging technologies demand our attention. It is our collective responsibility to ensure that the benefits of technology are harnessed for the greater good and that no one is left behind in the pursuit of progress.

As we end this book and embark on our individual and collective journeys, let us remember that the path ahead will not be without its obstacles. However, armed with the tools and insights shared in this book, we have the capacity to thrive amidst uncertainty, to leverage technology for the greater good, and to shape a future that reflects our deepest aspirations. The time is now. We cannot afford to remain complacent or indifferent to the transformative power of technology. Making technology an ally, continually learning, and being open to change is a responsibility we owe to ourselves, our communities, and future generations. Now is the time to be agents of change. We have to embark on this journey with courage, curiosity, and a commitment to continual growth. By doing so, we can create a world where humans and technology coexist harmoniously, where innovation flourishes, and where our collective potential knows no bounds. In this age of rapid technological change, the power lies within each of us to shape our destiny, contribute to the betterment of society, and create a future that leans into the endless possibilities of human potential. We can build a future that harnesses technology, continually learns, and embraces change - today, tomorrow, and for generations to come.

This start of the Fourth Industrial Revolution has bestowed upon us a world of immense possibilities and untapped potential. Each of us have the opportunity to shape the narrative of the digital age, to redefine the role of technology in our lives, and to create a future where humans and machines coexist symbiotically. Together, we can navigate the uncharted waters of the digital age and build a future that thrives on the fusion of human ingenuity and technological progress. A future where we survive our own innate biases and resistance to change, and one in which we can capitalize on the abundant opportunities ahead that allow us to thrive.

This is a future in which we can truly "**thurvive**".

A future in which we are the Human Inhabitants of this Digital Planet.

|| END ||

|| 01000101 01001110 01000100 ||

Bibliography

Ali, A. (2021, November 10). *From Amazon to Zoom: What Happens in an Internet Minute In 2021?* Retrieved from Visual Capitalist: https://www.visualcapitalist.com/from-amazon-to-zoom-what-happens-in-an-internet-minute-in-2021/

Alton, L. (2018, October). *The Evolution From Work-Life Balance to Work-Life Integration.* Retrieved from ADP: https://www.adp.com/spark/articles/2018/10/the-evolution-from-work-life-balance-to-work-life-integration.aspx

Amazon. (2021). *Amazon Astro.* Retrieved from Amazon.com: https://www.amazon.com/Introducing-Amazon-Astro/dp/B078NSDFSB

Amazon. (2021). *The top 500 sites on the web.* Retrieved from Alexa, An Amazon Company: https://www.alexa.com/topsites

Ang, C. (2021). *Ranking U.S. Generations on Their Power and Influence Over Society.* Visual Capitalist.

ARCHAMBAULT, M. (2021, December 10). *25 First Photos from the History of Photography.* Retrieved from PetaPixel: https://petapixel.com/2021/12/10/first-photos-from-the-history-of-photography/

BLS. (2011, March 01). *Unemployment rates in 2010.* Retrieved from US Bureau of Labor Statistics: https://www.bls.gov/opub/ted/2011/ted_20110301.htm?view_full

Bluetooth-SIG. (n.d.). *Bluetooth.* Retrieved from Bluetooth Special Interest Group: https://www.bluetooth.com/

Britannica, T. E. (2021, July 21). *https://www.britannica.com/event/Industrial-Revolution.* Retrieved from Encyclopedia Britannica: https://www.britannica.com/event/Industrial-Revolution

Britannica, T. E. (n.d.). *The Roman Empire.* Retrieved from Encyclopedia Britannica: https://www.britannica.com/place/Roman-Empire

Broad Institute, T. (2021). *Questions and Answers About CRISPR.* Retrieved from The Broad Institute: https://www.broadinstitute.org/what-broad/areas-focus/project-spotlight/questions-and-answers-about-crispr

Brown, S. (2021, October 2). *Amazon Astro household robot.* Retrieved from CNET: https://www.cnet.com/home/smart-home/amazon-astro-household-robot-everything-to-know-about-price-privacy-battery-and-more/

California History - State Symbols. (n.d.). Retrieved from California State Library: https://www.library.ca.gov/california-history/state-symbols/

Chapman, L. (2021, August 11). *Uber-Backed Joby Surges on the Promise of Air Taxis by 2024.* Retrieved from Bloomberg: https://www.bloomberg.com/news/articles/2021-08-11/uber-backed-joby-goes-public-on-the-promise-of-air-taxis-by-2024

Cheng, S. (2017, February 17). *For a little perspective ….* Retrieved from CUNY: https://openlab.citytech.cuny.edu/neh-digitech/2016/02/17/for-a-little-perspective/

CIE. (1999, April 05). *M-Systems Patents USB Flash Drive.* Retrieved from Center for Israel Education: https://israeled.org/m-systems-patents-usb-flash-drive/

Clarke, A. C. (1962). Hazards of Prophecy: The Failure of Imagination. In A. C. Clarke, *Profiles of the Future.*

Curry, D. (2022, January 11). *Video Streaming App Revenue and Usage Statistics (2022).* Retrieved from Business of Apps: https://www.businessofapps.com/data/video-streaming-app-market/

Darwin, C. (1859). *On the Origin of Species.*

Dastin, J. (2018, October 10). *Amazon scraps secret AI recruiting tool that showed bias against women.* Retrieved from Reuters: https://www.reuters.com/article/us-amazon-com-jobs-automation-insight/amazon-scraps-secret-ai-recruiting-tool-that-showed-bias-against-women-idUSKCN1MK08G

Desoutter. (n.d.). *Industrial Revolution - From Industry 1.0 to Industry 4.0.* Retrieved from Desoutter Industrial Tools: https://www.desouttertools.com/industry-4-0/news/503/industrial-revolution-from-industry-1-0-to-industry-4-0

Deutschman, A. (2007). *Change or Die: The Three Keys to Change at Work and in Life.* Harper Business.

Digital.net. (2011, 02). *Digital.net.* Retrieved from Netflix vs Blockbuster Revenues: http://go-digital.net/blog/wp-content/uploads/2011/02/netflix-vs-blockbuster-revenues.gif

Displays2go. (2021, November 29). *Touchscreens - Past, Present, Future.* Retrieved from Displays2Go: https://www.displays2go.com/Article/Touchscreens-Past-Present-Future-87#:~:text=The%20first%20touchscreen%20was%20invented,a%20diagram%20of%20the%20design.

Editors, H. (2009, November 16). *A thousand pioneers head West as part of the Great Emigration.* Retrieved from History.com: http://www.history.com/this-day-in-history/a-thousand-pioneers-head-west-on-the-oregon-trail

Encyclopedia. (n.d.). *The Invention Of The Fax Machine.* Retrieved from Encyclopedia.com: The Invention Of The Fax Machine

Eplett, L. (2015, June 16). *Ain't Nobody Got Tine For That! The Invention and Evolution of the Fork.* Retrieved from Scientific American: https://blogs.scientificamerican.com/food-matters/ain-t-nobody-got-tine-for-that-the-invention-and-evolution-of-the-fork/

Facebook. (2021, April 28). *Facebook First Quarter 2021 Results.* Retrieved from Facebook Investor Relations: https://investor.fb.com/investor-news/press-release-details/2021/Facebook-Reports-First-Quarter-2021-Results/default.aspx

Faruqui, A., & Eakin, K. (2000). *Pricing in Competitive Electricity Markets.* Springer Science & Business Media.

Frankel, M. (2017, July 29). *9 Baby-Boomer Statistics That Will Blow You Away.* Retrieved from The Motley Fool: https://www.fool.com/retirement/2017/07/29/9-baby-boomer-statistics-that-will-blow-you-away.aspx

Future of Life Institute. (Retrieved 2023, May 02). *About us.* Retrieved from Future of Life Institute: https://futureoflife.org/about-us/

Gallagher, W. (2022, January 25). *Original 'iBeer' iPhone app made creators $20,000 a day.* Retrieved from Appinsider: https://appleinsider.com/articles/22/01/25/original-ibeer-iphone-app-made-creators-20000-a-day

Gartner. (n.d.). *Gartner Hype Cycle.* Retrieved from Gartner: https://www.gartner.com/en/research/methodologies/gartner-hype-cycle

Golder, A. (2016, December 10). *23 Mind Blowing Facts About Time.* Retrieved from BuzzFeed: https://www.buzzfeed.com/andyneuenschwander/23-mind-blowing-facts-about-time-thatll-make-you-say-whoa

Google. (n.d.). *Google Quantum AI.* Retrieved from Google Quantum AI: https://quantumai.google/

Heath, A. (2021, October 25). *FACEBOOK'S LOST GENERATION.* Retrieved from The Verge: https://www.theverge.com/22743744/facebook-teen-usage-decline-frances-haugen-leaks

Heaven, W. D. (2023, May 02). *Geoffrey Hinton tells us why he's now scared of the tech he helped build.* Retrieved from MIT Technology Review: https://www.technologyreview.com/2023/05/02/1072528/geoffrey-hinton-google-why-scared-ai/

Heimlich, R. (2010). *Baby Boomers Retire.* PEW RESEARCH CENTER.

Hill, J. (2017, March 27). *The smart home: a glossary guide for the perplexed.* Retrieved from T3: https://www.t3.com/features/the-smart-home-guide

Hootsuite. (2020). *Digital 2020.* Vancouver, BC, Canada: Hootsuite.

IMDB. (2017, July 17). *IMDB All Jurassic Park Movies.* Retrieved from IMDB: https://www.imdb.com/list/ls068935685/

IMDB. (n.d.). *Cosmos* . Retrieved from IMDB: https://www.imdb.com/title/tt0081846/

James, J. (2021, October 04). *What 'Data Never Sleeps 9.0' Proves About the Pandemic.* Retrieved from Domo: https://www.domo.com/blog/what-data-never-sleeps-9-0-proves-about-the-pandemic/

Juma, C. (2016). *Innovation and Its Enemies: Why People Resist New Technologies.* Oxford University Press.

Kendall, G. (2019, July 01). *Would your mobile phone be powerful enough to get you to the moon?* Retrieved from The Conversation: https://theconversation.com/would-your-mobile-phone-be-powerful-enough-to-get-you-to-the-moon-115933

Klein, E. (1915, September 21). *Lawrence Daily Journal-World Newspaper.* Retrieved from Lawrence Daily Journal-World Newspaper.

Kleinman, Z., & Vallance, C. (2023, May 02). *AI 'godfather' Geoffrey Hinton warns of dangers as he quits Google.* Retrieved from BBC: https://www.bbc.com/news/world-us-canada-65452940

Klok, C. H. (2020, February 17). *The messy, secretive reality behind OpenAI's bid to save the world.* Retrieved from MIT Technology Review: https://www.technologyreview.com/2020/02/17/844721/ai-openai-moonshot-elon-musk-sam-altman-greg-brockman-messy-secretive-reality/

Krulwich, R. (2010, October 25). *First Photo Of A Human Being Ever?* Retrieved from National Public Radio (NPR): https://www.npr.org/sections/krulwich/2011/03/31/130754296/first-photo-of-a-human-being-ever?ps=cprs

Kurzweil, R., & Meyer, C. (2003, May 2). *Understanding the Accelerating Rate of Change.* Retrieved from Kurzweil:

https://www.kurzweilai.net/understanding-the-accelerating-rate-of-change

Lee, E. (2021). *2021 Worldwide Image Capture Forecast: 2020 – 2025.* Rise Above Research.

Lewis, L. (2021, April 13). *Infographic: What Happens In An Internet Minute 2021.* Retrieved from Merge - All Access: https://www.allaccess.com/merge/archive/32972/infographic-what-happens-in-an-internet-minute

Liberatore, S. (2016, December 16). *What the internet looked like in 1973: Tiny map shows the ENTIRE network was made up of just 42 computers.* Retrieved from Daily Mail: https://www.dailymail.co.uk/sciencetech/article-4038292/What-internet-looked-like-1973-Tiny-map-shows-ENTIRE-network-just-42-computers.html

Lisa, A. (2021, October 21). *Jobs that might not exist in 50 years.* Retrieved from Stacker: https://stacker.com/careers/jobs-might-not-exist-50-years

McKinsey & Company. (2017, September 07). *Human + machine: A new era of automation in manufacturing.* Retrieved from McKinsey & Company: https://www.mckinsey.com/capabilities/operations/our-insights/human-plus-machine-a-new-era-of-automation-in-manufacturing

McKinsey Global Institute. (2015, December 01). *Digital America: A tale of the haves and have-mores.* Retrieved from McKinsey & Company: https://www.mckinsey.com/industries/technology-media-and-telecommunications/our-insights/digital-america-a-tale-of-the-haves-and-have-mores

McKinsey Global Institute. (2017). *Jobs lost, jobs gained: What the future of work will mean for jobs, skills, and wages.* McKinsey & Company.

McSweeney, K. (2020, January 31). *NOW Powered by Northrop Grumman.* Retrieved from Northrop Grumman: https://now.northropgrumman.com/ray-kurzweil-predictions-persist-turns-70/

Megginson, L. (1963). Lessons from Europe for American Business. *Southwestern Social Science Quarterly*, 44(1): 3-13, at p. 4.

Metz, C., & Weise, K. (2023, January 12). *Microsoft Bets Big on the Creator of ChatGPT in Race to Dominate A.I.* Retrieved from The New York Times: https://www.nytimes.com/2023/01/12/technology/microsoft-openai-chatgpt.html

Meyersohn, N. (2018, March 06). *Why Domino's is winning the pizza wars.* Retrieved from CNN Money: https://money.cnn.com/2018/03/06/news/companies/dominos-pizza-hut-papa-johns/index.html

Naftulin, J. (2016, Juy 13). *Here's how many times we touch our phones every day.* Retrieved from Business Insider: https://www.businessinsider.com/dscout-research-people-touch-cell-phones-2617-times-a-day-2016-7

NASA. (n.d.). *Europa Clipper Mission.* Retrieved from NASA: https://www.jpl.nasa.gov/missions/europa-clipper

Neufield, D. (2021, November 08). *There are 1.8 billion millennials on earth. Here's where they live.* Retrieved from World Economic Forum : https://www.weforum.org/agenda/2021/11/millennials-world-regional-breakdown/

Neuman, S. (2021, September 14). *Scientists Say They Could Bring Back Woolly Mammoths. But Maybe They Shouldn't.* Retrieved from National Public Radio (NPR.org): https://www.npr.org/2021/09/14/1036884561/dna-resurrection-jurassic-park-woolly-mammoth

Niiler, E. (2019, January 25). *How the Second Industrial Revolution Changed Americans' Lives.* Retrieved from The History Channel: https://www.history.com/news/second-industrial-revolution-advances

Nikkei. (2021, July 09). *JAL aims to launch flying car business in fiscal 2025.* Retrieved from Nikkei Asia:

https://asia.nikkei.com/Business/Technology/JAL-aims-to-launch-flying-car-business-in-fiscal-2025

Nokiamob. (2017, December 11). *Can Anyone Catch Nokia*. Retrieved from Nokiamob: https://nokiamob.net/2017/11/12/forbes-in-2007-can-anyone-catch-nokia/

NPR. (2021). *NPR Special Series - The CRISPR Revolution*. Retrieved from National Public Radio (NPR): https://www.npr.org/series/773368439/the-crispr-revolution

NYT. (1973, 09 13). *Mrs. Marjorie Merriweather Post Is Dead at 86*. Retrieved from The New York Times: https://www.nytimes.com/1973/09/13/archives/mrs-marjorie-merriweather-post-is-dead-at-86-a-rich-working-woman.html

NYT. (2022, January 3). *NYT Technology*. Retrieved from The New York Times: https://www.nytimes.com/2022/01/03/technology/personaltech/blackberry-devices-stop-working.html

OpenAI. (2021). *OpenAI Codex*. Retrieved from OpenAI Codex: https://openai.com/

Overly, S. (2016, July 21). *Humans once opposed coffee and refrigeration. Here's why we often hate new stuff.* Retrieved from The Washington Post: https://www.washingtonpost.com/news/innovations/wp/2016/07/21/humans-once-opposed-coffee-and-refrigeration-heres-why-we-often-hate-new-stuff/

Pande, P. (2021, July 12). *JAL Looks At eVTOL Operations From 2025*. Retrieved from Simple Flying: https://simpleflying.com/jal-evtol-2025/

Pantic, N. (2021). *How Many Photos Will Be Taken in 2021?* Retrieved from mylio: https://blog.mylio.com/how-many-photos-will-be-taken-in-2021-stats/

Pew. (2015). *Labor Force Composition by Generation.* PEW RESEARCH CENTER.

Phillpott, S. (2021, May 17). *15 Disappearing Jobs that Won't Exist in 2030.* Retrieved from CareerAddict: https://www.careeraddict.com/disappearing-jobs

Profit&. (2019, September 20). *7 REAL-LIFE EXAMPLES OF SUCCESSFUL CHANGE MANAGEMENT IN BUSINESS.* Retrieved from Profit&: https://insights.profitand.com/blog/real-life-examples-of-successful-change-management-in-business

Purdue. (2020). *Generational Differences in the Workplace.* Retrieved from Purdue University Global: https://www.purdueglobal.edu/education-partnerships/generational-workforce-differences-infographic/

Reddy, C. (2017, October 05). *Kurzweil Claims That the Singularity Will Happen by 2045.* Retrieved from Futurism.com: https://futurism.com/kurzweil-claims-that-the-singularity-will-happen-by-2045

Richards, A. (2012, May 18). *Clarence Birdseye And His Fantastic Frozen Food Machine.* Retrieved from National Public Radio: https://www.npr.org/sections/thesalt/2012/05/18/152743718/clarence-birdseye-and-his-fantastic-frozen-food-machine

Roser, M., Ortiz-Ospina, E., & Ritchie, H. (2019). *Life Expectancy (United Nations, Our World In Data).* United Nations.

Schwab, K. (2015, December 12). *The Fourth Industrial Revolution - What It Means and How to Respond.* Retrieved from Foreign Affairs: https://www.foreignaffairs.com/articles/2015-12-12/fourth-industrial-revolution

Schwab, K. (2017). *The Fourth Industrial Revolutuion.* Currency.

Scientist, N. (2021). *New Scientist.* Retrieved from New Scientist: https://www.newscientist.com/definition/what-is-crispr/

Snider, M., & Molina, B. (2021, November 10). *Everyone wants to own the metaverse including Facebook and Microsoft. But what exactly is it?* Retrieved from USA Today:

https://www.usatoday.com/story/tech/2021/11/10/metaverse-what-is-it-explained-facebook-microsoft-meta-vr/6337635001/

Stein, R. (2021, June 26). *National Public Radio - All Things Considered.* Retrieved from National Public Radio (NPR): https://www.npr.org/sections/health-shots/2021/06/26/1009817539/he-inherited-a-devastating-disease-a-crispr-gene-editing-breakthrough-stopped-it

study.com. (2014, February 06). *The Second Industrial Revolution: Timeline & Inventions.* Retrieved from Study.com: https://study.com/academy/lesson/the-second-industrial-revolution-timeline-inventions.html

Surowiecki, J. (2013, September 03). *Where Nokia Went Wrong.* Retrieved from The New Yorker: https://www.newyorker.com/business/currency/where-nokia-went-wrong

Terrell, K. (2022, June 02). *10 Jobs That Are Disappearing Fast.* Retrieved from AARP: https://www.aarp.org/work/careers/disappearing-jobs/

TKSST. (2012). *Carl Sagan's Cosmic Calendar.* Retrieved from The Kid Should See This - Smart videos for curious minds of all ages: https://thekidshouldseethis.com/post/cosmic-calendar-carl-sagan-video

Tocci, M. (2020). *What Pleasures Would You Sacrifice to Keep Your Smartphone?* SimpleTexting.

Vega, N. (2021, December 27). *Microsoft's market cap grew more than $800 billion in 2021—here's how it compares to the most valuable companies in the world.* Retrieved from CNBC: https://www.cnbc.com/2021/12/27/how-much-the-biggest-companies-grew-in-2021.html

Wall, M. (2021, July 7). *Enceladus.* Retrieved from Space.com: https://www.space.com/methane-plume-enceladus-possible-sign-alien-life

Weidner, J. B. (2021, August 8). *How and Why Google Glass Failed.* Retrieved from Investopedia:

https://www.investopedia.com/articles/investing/052115/how-why-google-glass-failed.asp

Weitz, A. (2017). *The Camera Phone is 20 Years Old!* Retrieved from B&H: https://www.bhphotovideo.com/explora/photography/features/camera-phone-20-years-old

Wikipedia. (2022). *iPhone.* Retrieved from Wikipedia: https://en.wikipedia.org/wiki/IPhone

Wikipedia. (n.d.). *USB.* Retrieved from Wikipedia: https://en.wikipedia.org/wiki/USB

Wikipedia. (n.d.). *Wikipedia.* Retrieved from Clarence Birdseye: https://en.wikipedia.org/wiki/Clarence_Birdseye

Wilkins, A. (2012, March 25). *The last mammoths died out just 3600 years ago...but they should have survived.* Retrieved from Gizmodo: https://gizmodo.com/the-last-mammoths-died-out-just-3600-years-ago-but-th-5896262

Wilkinson, D. (2020). *2020 SMS Marketing Report.* SimpleTexting.

Witten, S. (2019, November 08). *The death of the DVD: Why sales dropped more than 86% in 13 years.* Retrieved from CNBC: https://www.cnbc.com/2019/11/08/the-death-of-the-dvd-why-sales-dropped-more-than-86percent-in-13-years.html

Zurich. (2022, June 24). *How will Gen Z change the workplace?* Retrieved from Zurich: https://www.zurich.com/en/media/magazine/2022/how-will-gen-z-change-the-future-of-work#:~:text=Gen%20Z%20could%20change%20jobs,a%20survey%20of%20U.S.%20students.

About the Author

Kartik Sakthivel PhD serves as the Vice President and global Chief Information Officer for LIMRA LOMA and LL Global, a non-profit worldwide trade association for the financial services industry. Kartik has had a vast amount of leadership and management experience across multiple industries. He has held leadership roles in companies of all sizes, ranging from small and midsize firms to Fortune 100 organizations.

A technology practitioner, Kartik has been urging industries and workers within the industries to adopt and adapt to the massive digital disruption that is the hallmark of the early 21st century. Believing that encouraging industries and people to continually invent and reinvent themselves is an act of consummate leadership, Kartik is on a mission to impress upon leaders how crucial they are to the success of their organizations and their people. With an immutable belief that leadership is not about a title, but how you serve your teams, Kartik has been coaching and mentoring new leaders on how they can start the process of leaving their leadership legacy.

Kartik holds a PhD in Computer and Information Science (with a focus on Information Quality and a dissertation on best practices for AI) and a Graduate Certificate in Information Quality from University of Arkansas at Little Rock, a Master's Degree in Computer Information Systems/IT, and a Master of Business Administration degree from Southern New Hampshire University.

Born in Mumbai (Bombay) India, Kartik has been a resident of Dover, New Hampshire in the United States of America for his entire adult life.

Digital Planet Human Inhabitants

Kartik Sakthivel

Made in the USA
Monee, IL
04 November 2026

67385374R00174